# CARRYING CAPACITY OF MUSSOORIE

## About the Author

**Dr. H. Ramachandran** is currently the Director of the Institute of Applied Manpower Research and was earlier Executive Director of the National Society for Promotion of Development Administration, Research and Training, Mussoorie. Prior to this he was associated with the Lal Bahadur Shastri National Academy of Administration, Mussoorie, the Delhi School of Economics and the Institute for Social and Economic Change, Bangalore. He has a number of books and papers in the field of urban and rural development to his credit.

The Supreme Court Monitoring Committee was constituted by the Hon'ble Supreme Court of India under a judgement of 30.8.1988 in a Public Interest Litigation Case, for monitoring the environment issues in Doon Valley.

# CARRYING CAPACITY OF MUSSOORIE

DR. H. RAMACHANDRAN
DR. NIRA RAMACHANDRAN *et al.*

*Published for*

**Supreme Court Monitoring Committee,**
**Dehradun-248006**

*by*

**CONCEPT PUBLISHING COMPANY, NEW DELHI-110059**

ISBN 81-7022-926-X

First Published 2001



*Published and Printed by*
Ashok Kumar Mittal
**Concept Publishing Company**
A/15-16, Commercial Block, Mohan Garden,
New Delhi-110059 (India)

*Phones*: 5648039, 5649024
*Fax*: 091–(11)–5648053
*E.mail*: publishing@conceptpub.com

# Preface

The urban centres in hill areas have in recent times attracted investments mainly in the construction sector. Much of this happens to be of speculative and commercial nature. This has obviously marginalised the resident population, particularly, with reference to land markets. Some of these towns also attract large numbers of tourists, who put additional pressure on urban infrastructure. At the same time, the tourists also happen to generate employment for the resident population and the city earns a substantial income. Both the economic and environmental costs of creating infrastructure are high. There is, therefore, the crucial question of carrying capacity and sustainability of the growth of urban centres in hill regions.

The study on Carrying Capacity was sponsored by the Supreme Court Monitoring Committee in April 1997, specifically to establish benchmarks on important parameters and pressure on infrastructure.

The study team included Dr. H. Ramachandran, Dr. Nira Ramachandran, Shri Padam Vir Singh, Ms. Seema Joshi, Shri A.W. Khan, Ms. Saloni Goal, Shri R.K. Bharadwaj and Ms. Madhuri Negi. A number of people have helped in the process of the study. We would like to specifically acknowledge assistance from Shri Jot Singh, Chairman, Municipal Council, Mussoorie, elected representatives of the Council, various departments of the Municipality, school authorities, officials of the government and other organizations, residents—particularly those who attended the meetings and gave their suggestions, the members of the Hotel Owners' Association, the National Remote Sensing Agency, the Indian Institute of Remote Sensing, and the officials of the MDDA.

Shri B. S. Baswan, Director, LBSNAA and Dr. V.K. Agnihotri, Joint Director, LBSNAA have been constantly monitoring the progress of the study, which helped the completion of the study within the stipulated time.

A draft report was circulated in April 1998 and the results of the study were presented and discussed in an open session in May 1998. All officials and non-officials who had taken part in the initial participatory meetings were invited and over forty participants attended the open session. This Revised Version of the report incorporates the suggestions and modifications proposed in the meeting.

Mussoorie
July 1998

**Dr. H. Ramachandran**
**Dr. Nira Ramachandran *et al.***
National Society for Promotion of
Development Administration,
Research and Training,
LBSNAA

# Contents


| | | |
|---|---|---|
| *Preface* | | 5 |
| *List of Tables* | | 8 |
| *List of Maps and Illustrations* | | 10 |
| 1. | Objectives, Methodology and Data Base | 11 |
| 2. | Problems and Public Interventions : An Analysis of Residents' Perception | 22 |
| 3. | On Measuring Carrying Capacity of Mussoorie | 35 |
| 4. | The Limits to City Expansion : The Use of Land in the Mussoorie Hills | 60 |
| 5. | Concluding Observations | 81 |
| | **Annexures** | |
| 1. | Water Supply and Sewerage System in Mussoorie | 85 |
| 2. | Waste Management in Mussoorie : A Case Study | 97 |
| | **Appendices** | |
| I. | Civic Services and Infrastructure Questionnaire | 112 |
| II. | Questionnaire–Waste Management | 122 |
| III. | Civic Services and Infrastructure | 126 |
| IV. | Tourist Inflow in Mussoorie | 127 |
| V. | Month-wise Variation in Population Pressure–1997 | 128 |
| VI. | Per Capita Waste Generation by Permanent Population | 139 |
| *Index* | | 142 |

# List of Tables


1.1 Mussoorie Municipality–Annual Accounts (Income) 20
1.2 Mussoorie Municipality–Annual Accounts (Expenditure) 21
2.1 Prioritisation of Problems by Various Interest Groups 23
2.2 Distribution of Residents' Response to Adequacy of Infrastructure by Educational and Income Levels 24
2.3 Distribution of Residents' Response to Adequacy of Water and Power Supply 25
2.4 Distribution of Residents' Response to Adequacy of Civic Services 26
2.5 Distribution of Residents' Response to Adequacy of Transport Services 27
2.6 Distribution of Residents' Response to Adequacy of Education and Health Facility 28
2.7 Distribution of Residents' Response to City Development Housing 29
2.8 Distribution of Residents' Perception of Pollution 30
2.9 Major Indicative Solutions Offered by Residents 31
3.1 Census Population of Municipal Board and Cantonment (1981 and 1991) 35
3.2 Population Estimates for Landour Cantonment Area 36
3.3 Mussoorie Urban Agglomeration–Population Growth (1901-91) 40
3.4 Population Projections for Mussoorie U.A. 1997-2021 43
3.5 Tourist In-flow to Mussoorie (1990-96) 45
3.6 Projected Water Demand @ 150 lpcd (Scenario - I) 46
3.7 Projected Water Demand @ 150 lpcd (Scenario - II) 47
3.8 Estimates of Waste Generation in Mussoorie (Scenario - I) 49

3.9 Estimates of Waste Generation in Mussoorie (Scenario-II) 51
3.10 Beds Available for Tourists in Mussoorie 54
3.11 Tourist Flow Per Day – Estimates (Mussoorie) 54
3.12 Availability of Hospital Beds in Mussoorie (1997) 57
3.13 Requirement of Hospital Beds (Scenario - I) 58
3.14 Requirement of Hospital Beds (Scenario - II) 58
3.15 Growth in Number of Electricity Connections and Consumption in Mussoorie 59
4.1 Area and Percentage of Land in Mussoorie by Slope Categories 64
4.2 Change in Landuse in Mussoorie (1990-97) 68
4.3 Use of Land in Mussoorie by area Under Various Categories of Slopes, 1990 72
4.4 Use of Land in Mussoorie–Percent Area under Various Land Use Categories by Degree of Slope 76
4.5 Mussoorie Land Conversion Matrix–1990-97 80

# List of Maps and Illustrations


| | | |
|---|---|---|
| 1A | Mussoorie–Space Relations | 14 |
| 1B | Mussoorie–Municipal Wards | 15 |
| 2 | Mussoorie–Ranking of Problems by Residents (First Ranking Problems) | 32 |
| 3 | Mussoorie–Ranking of Problems by Residents (Second Ranking Problems) | 33 |
| 4 | Mussoorie–Population Growth (1901-91) | 41 |
| 5 | Mussoorie–Spatial Distribution of Dustbins | 56 |
| 6A | Mussoorie–Landuse (1990) | 62 |
| 6B | Mussoorie–Landuse (1997) | 63 |
| 7 | Mussoorie–Slope Analysis | 66 |
| 8A | Mussoorie–Percent Built-up Area (1990) and (1997) | 67 |
| 8B | Mussoorie–Built-up Area (Addition between 1990-1997) | 67 |
| 9 | Mussoorie–Percent Area Under Dense Forest (1990 qne 1997) | 70 |
| 10 | Mussoorie–Percent Area Under Open Forest (1990 and 1997) | 71 |
| 11A | Percentage Landuse Under Various Slope Classes (1990 and 1997) | 74 |
| 11B | Percentage Landuse Under Various Slope Classes, 1997 | 75 |
| 12A | Mussoorie: The Use of Land: Area Below 30 Degree Slope–Build-up Area | 78 |
| 12B | Mussoorie–The Use of Land: Area Below 30 Degree Slope–Barren Land | 78 |
| 12C | Mussoorie–The Use of Land: Area Below 30 Degree Slope–Cultivated Area | 79 |
| 12D | Mussoorie–The Use of Land: Area Below 30 Degree Slope–Land Forests | 79 |

# 1

# Objectives, Methodology and Data Base

## 1.1 Background

Located at an altitude of about 2,000 meters in the lower Himalayan ranges of Dehradun district, the hill resort of Mussoorie covers an area of 67.75 square kilometres (Figures 1A and 1B). The Census of India (1991) records a resident population of 29,629 in the Mussoorie Urban Agglomeration (U.A.), which includes the Municipality and the Cantonment area.

Many believe that this is an underestimate since enumeration was done in the month of February (Sunrise of March 1 as the reference moment) when a significant proportion of the city population that normally moves down to the plains during the winter had not returned. Besides, often the population figures quoted ignore a substantial number of children studying in the boarding schools of the city, and the population in the Tibetan settlements, etc. While the Census figures suggest a fluctuating, but sedate growth of population of the city, the tourist inflow has been rising far more rapidly over the years, wherein during the summers the floating population can be as high as twice the permanent population. Public resource allocation for city infrastructure development generally takes into account the number of permanent population. Like all resort towns, the seasonal influx of tourists overloads an already overburdened city infrastructure of Mussoorie. As the mainstay of the town is the tourist industry, any plan for sustaining the town, must, of necessity, provide built-in the demand for tourist facility of which real estate factor is a component. Such a plan must also project future demands of tourism and lay down guidelines for encouraging this industry while sustaining the resource base and environment. It is also to be noted that there is a circular causation between tourism and the environment–unrestrained flow of tourists beyond the city's capacity, with the concomitant strain on the infrastructure and the environment—which results in a deteriorating environment and a reduction in the tourist flow. The available information base is rather weak, even with reference to such basic data as population figures.

Other than a few government and quasi-government organisations and a number of schools, which together have over 6,000 boarders, the main economic base of Mussoorie is the tourist sector. Consequently, the service sector is the mainstay of the city. Precisely because of this, this small town with limited developable land base–owing to its hill terrain–records about 140 hotels, *dharamshalas* and guest houses which together have a bed capacity for accommodating nearly 10,000 people.

The development of the city comes under the jurisdiction of the Mussoorie-Dehradun Development Authority (MDDA) which was created through a Government Notification (No.225/XI-5-84-19(1)-DA.84, dated October 29 1984, in exercise of the powers under section 3 of the Uttar Pradesh Urban Planning and Development Act, 1973) with a view to developing the area in a planned manner. The Development by-laws of the MDDA 1985, apply to development activities and existing buildings within the development area of the MDDA. All development and re-development permits are thus to be obtained from the MDDA. Permission is necessary to erect, re-erect, demolish, and make any material change or alteration in a building or part thereof within the development area. However, no permission is required for the following alterations–provision or closure of a window or a door or ventilator not opening towards other's property, providing inter-communication doors, providing partitions, gardening, white- washing, painting, re-tiling, plastering, re-flooring and construction of sun-shades. Similarly, no permits are required for carrying out works by departments of the Central or State Government or any local body for the purpose of inspection, repairing or renewing any sewers, mains, pipes, cables or other apparatus.

The increase in the built-up area of the city has been phenomenal, as we will see later (Chapter 4), between 1990 and 1997, and the shrinking of forest cover is also striking. Meanwhile a Monitoring Committee was constituted by the Honourable Supreme Court of India with reference to Doon Valley development.

The Honourable Supreme Court in their interim order delivered on July 10, 1996, on a writ by the Supreme Court Monitoring Committee versus MDDA, stated that constructions which have not begun at site and have not proceeded beyond the plinth level shall not be permitted to be started till the stand of the State of Uttar Pradesh and the Union of India on the applicability or otherwise of the provisions of the Forest Conservation Act, 1980, and the rules made thereunder is made clear. The Honourable Court subsequently clarified on December 15, 1997, that the order dated July 10, 1996, is not to be construed as permitting the continuance of construction which had proceeded beyond the plinth level on the date of the order. In the order dated November 29, 1996, the direction given to the Central Government was to consider the cases for grant of *ex post facto* approval. This needs to be read along with the other orders made by the court in these proceedings as well as in another matter (writ petition (C) No. 202 of 1995–T.N. Godavarman *Vs.* Union of India & Ors. SCALE 1996(9) 269).

It is in this background that a study was proposed and funded by the Supreme Court Monitoring Committee to assess the environmental status of Mussoorie in the larger context of urban and non-urban land use, and the carrying capacity of urban infrastructure.

## 1.2 Objectives of the Study

The specific objectives of the study may be stated as follows:

(a) to bring out the current rate of conversion of open space to built-up space and to review the current procedures and rules relating to city land use;

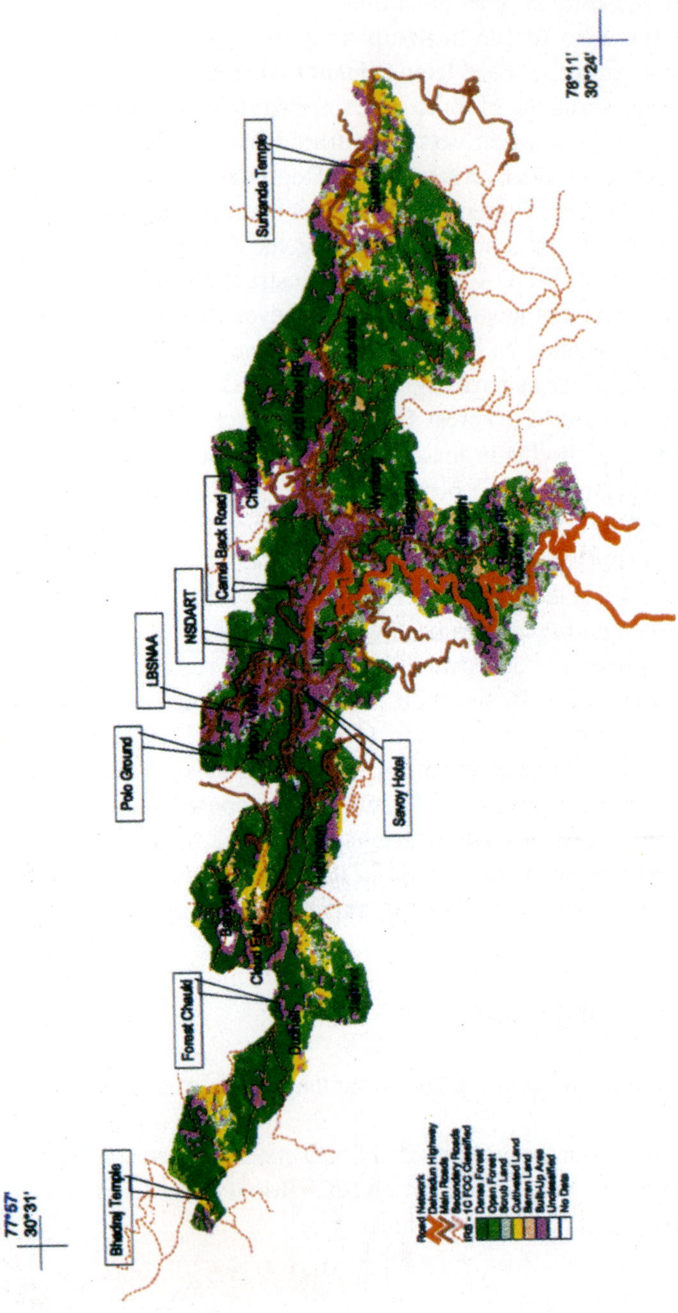


**Fig. 1A : Mussoorie–Space Relations**

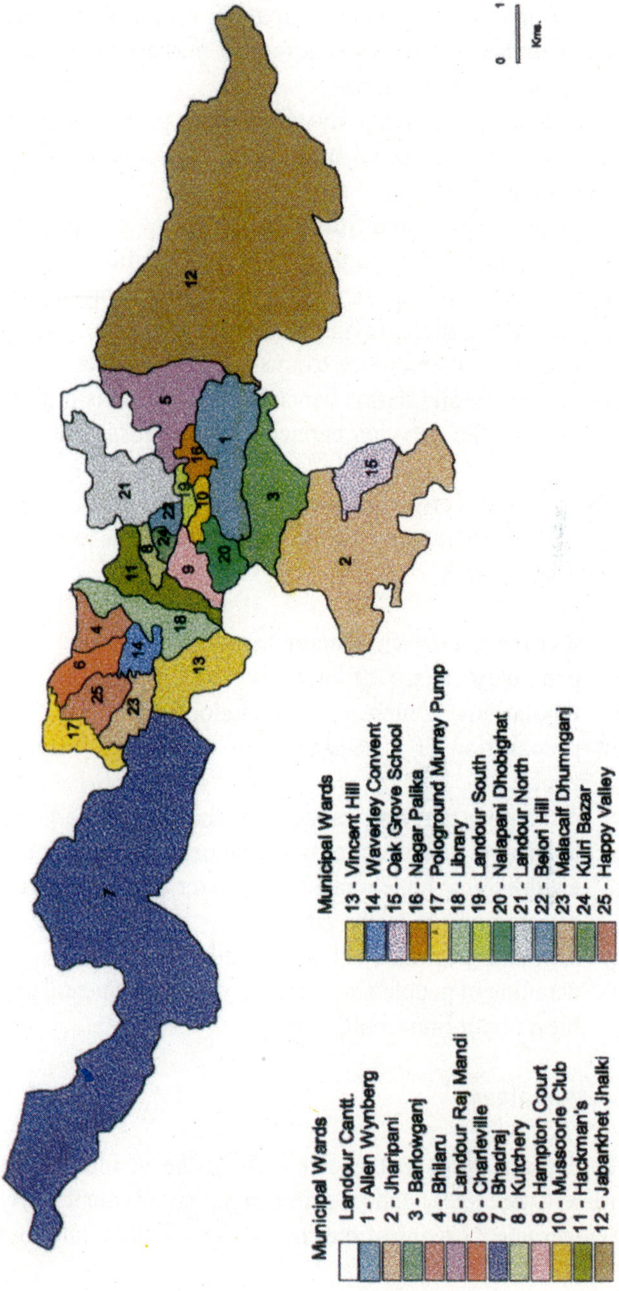


**Fig. 1B : Mussoorie–Municipal Wards**

(b) to prepare a detailed land use map including vertical spread and intensity of land use in areas with different slopes;
(c) to bring out the problems as perceived by different segments of residents;
(d) to analyse the ways in which these problems need to be tackled as perceived by various segments of the resident population;
(e) to undertake quantitative analysis of primary data of variables relating to the problems identified in (c) above;
(f) to develop indicators for monitoring carrying capacity and estimating the carrying capacity of the Mussoorie U.A. with special reference to available infrastructure; and
(g) to identify and define benchmarks of various indicators to monitor the carrying capacity of the Mussoorie hills.

The study was originally proposed in two phases. This report, based on the work relating to the first phase, was to address the following objectives:

(i) preparing a detailed land use map;
(ii) preparing a set of map overlays based on building regulations, zoning norms and slope stability;
(iii) preparation of a detailed map identifying hazard-prone zones;
(iv) estimating carrying capacity of the Mussoorie U. A. with particular reference to available urban infrastructure;
(v) establishing benchmarks for water and air pollution at selected points in the city;
(vi) estimating demand for water and power; and
(vii) detailing of people's perceptions on environmental problems based on a household survey.

## 1.3 Methodology

The key role players in the city include the residents, the local government agencies, schools and other government as well as quasi-government organisations, hotel owners, real estate and other businessmen. Since a study of this nature is geared ultimately to lead to action, it is necessary to make it as participatory as possible. As a result, emphasis was laid on group meetings with various interest groups such as the elected representatives, officials of the government and quasi-government organisations, members of the Hotel Owners' Association and groups of residents in each ward of the city. These meetings concentrated on prioritisation of problems and possible solutions to the problems identified.

Two kinds of schedules–one concentrating on the residents' backgrounds along with their responses relating to civic services and infrastructure (Appendix 1) and the other on waste management and garbage disposal (Appendix 2)—were canvassed among sample residents and institutions including hotels. The former questionnaire was canvassed at the venue of group meetings in each ward (a total of 311 respondents–for details see Appendix 3) whereas the latter was canvassed at the randomly drawn sample households (225 samples). The questionnaire dealing with civic services and infrastructure was also canvassed at other meetings with interest groups such as elected representatives of the City Board (17 respondents), hotel owners (18 respondents) and officials of the government and quasi-government organisations (19 respondents). While the questionnaire dealing with waste management and garbage disposal has been used to estimate the quantum of waste generated, etc., the questionnaire on civic services and infrastructure was basically used to understand the perception of problems by groups of residents and interest groups. Data has been gathered from various sources such as the Census of India, City Board, MDDA, the Divisional Forest Office, UPESB, Jal Nigam, Indian Institute of Petroleum (Dehradun), Regional Transport Office (Dehradun), and Landour Cantonment Board (Mussoorie). It must, however, be noted that often, these secondary data are not in a form in which they could be used for the purpose at hand. As a result, a number of estimates had to be made. The procedures used for estimating various parameters are explained at appropriate places in the report. Data has also been generated from various topographical maps and IRS LISS 3 FCC imageries (for 1990 and 1997). The maps have been digitised and the imagery geo-referenced and analysed on a GIS platform, for the purpose of studying land use, slope and overlays of maps.

### 1.4 An Introduction to City Administration and Finances

Consequent upon the 74th Constitutional Amendment (1992), and the Uttar Pradesh *Nagar Palika* Act, elections to the Mussoorie Municipal Board were held in 1997. The Municipality is divided into 25 wards (Figure 1A). Each ward is represented in the Municipal Board by one elected representative, besides the Chairman who is directly elected by the voters in the city. On an average then, each elected representative in the city represents about 1,500 residents. The tenure of the previous elected body of the Mussoorie Municipal Board came to an end on January 18, 1994, and subsequently for three years from 1994 to March 1997, an administrator managed the city.

The Municipal Board meets once every month. The major items discussed in these meetings deal with people's problems, problems of maintenance of services in their jurisdiction and the budget. The Board has created five sub-committees from among their members–property taxation committee, garden committee, health committee, public works committee and finance committee. These committees make the budget proposals and the proposals are scrutinised by the finance committee and then approved by the Municipal Board.

Three key functionaries – the Executive Officer, the Health Officer and the Municipal Engineer – who come from various provincial services provide administrative support to the Municipal Board. They are supported by a number of Assistant Engineers/ SDOs, Sanitary Inspectors and other junior-level officials. Some of the other basic infrastructure such as drinking water, power, education, telephone, etc., are outside the Municipal Board's jurisdiction and come under various state level institutions such as *Jal Sansthan,* State Electricity Board, the Department of Basic Education etc.

The annual revenue and receipts of the Municipal Board are given in Tables 1.1 and 1.2. The total income of the Municipality during 1996-97 was about Rs. 2.67 crores (excluding opening balance), of which locally generated revenue was about 45 per cent, the remaining coming in as grant-in-aid and subsidy from the state government. Between 1993-94 and 1996-97 the proportion of grant-in-aid and subsidy from the state government to the total income has been fluctuating — 51 per cent, 40 per cent, 39 per cent and 57 per cent. The per capita annual Municipal expenditure works out to Rs. 753 if we take the permanent population of the Municipality, which comes down to Rs.623 if we include the floating population.

## 1.5 Organisation of the Report

The report is organised in five chapters, the first chapter setting the background, the objectives, the approach and the information base used in the study. The second chapter deals with the analysis of residents' perception of problems relating to urban infrastructure and services. This forms the basis of the third chapter, which deals with the quantitative assessment of population pressure on infrastructure including projections for the next 25 years. The fourth chapter deals with an analysis of land use with particular reference to expansion of built-up area over time in relation to the degree of slope of the land and attempts to work out the percentage of built-up area in different categories of slope. The concluding remarks are included in the fifth chapter.

After the completion of the study, the draft report was presented at an open session to which were invited all officials and non-officials who had taken part in the initial participatory meetings. Over 40 participants attended the open session where the findings of the study were discussed and reviewed in the context of local knowledge and experience and various suggestions were put forth. This revised version of the report incorporates the suggestions and the modifications proposed in the meetings.

**Table 1.1:** Mussoorie Municipality–Annual Account (Income)

| S.NO. | HEADS OF INCOME | 1993-94 | 1994-95 | 1995-96 | 1996-97 |
|---|---|---|---|---|---|
| 1. | By Balance in hand at close of last year | 10,66,074 | 22,27,450 | 17,59,628 | 14,31,337 |
| 2. | Taxes on the annual value of building and land | 21,17,759 | 35,51,955 | 36,95,764 | 50,31,511 |
| 3. | Conservancy Taxes | 10,10,426 | 15,78,844 | 19,55,154 | 12,33,733 |
| 4. | Other taxes–(Animals, Pounds, Vehicles–Slaughter and Show taxes) | 16,626 | 26,478 | 20,739 | 12,97,152 |
| 5. | Revenue derived from Municipal Property and Powers apart from taxation (Rents of lands, houses) | 38,73,354 | 31,31,879 | 46,11,868 | 46,63,498 |
| 6. | Total fee recovered–(Tramways, Registration, Copying, Licence & Miscellaneous) | 1,99,068 | 1,75,814 | 1,62,809 | 1,84,629 |
| 7. | Interest on Investments | 91,339 | 1,19,371 | — | 1,66,986 |
| 8. | Total Help and Grants | 1,00,86,229 | 72,74,440 | 81,63,300 | 1,19,70,650 |
| 9. | Miscellaneous Taxes | 21,06,570 | 21,36,869 | 21,38,268 | 22,31,928 |
| | Total Income of year excluding opening balance | 1,95,01,371 | 179,95,650 | 2,07,47,902 | 2,67,80,087 |
| | Other advances and deposits | 25,314 | 40,405 | 3,775 | 6,765 |
| | Grand Total (including opening balance) | 2,05,92,759 | 2,02,63,505 | 2,25,11,305 | 2,82,18,189 |

**Table 1.2:** Mussoorie Municipality - Annual Account (Expenditure)

| S.NO. | HEADS OF EXPENDITURE | 1993-94 | 1994-95 | 1995-96 | 1996-97 |
|---|---|---|---|---|---|
| 1. | General Administration & Collection Charges | 13,17,549 | 14,92,492 | 19,84,371 | 19,11,957 |
| 2. | Collection Charges | 20,49,442 | 21,60,753 | 23,71,905 | 25,43,509 |
| 3. | Public Safety | 1,65,640 | 2,57,973 | 4,37,139 | 3,67,536 |
| 4. | Public Health Convenience | | | | |
| a. | Drainage | - | - | 23,628 | - |
| b | Conservancy | 2,84,314 | 51,08,165 | 55,58,974 | 64,46,516 |
| c | Health & Hospital (Vaccinations etc.) | 75,88,82 | 7,76,007 | 8,64,676 | 3,55,080 |
| d | Others | 46,69,101 | 2,77,628 | 4,59,928 | - |
| 5. | Horticulture & Gardening | 3,87,213 | 3,21,098 | 3,44,091 | 4,60,430 |
| 6. | Public work (Establishments, Buildings, roads, Stores) | 31,84,564 | 24,44,178 | 39,96,655 | 4,930,615 |
| 7. | Other Expenditure | 26,73,138 | 24,83,918 | 30,97,028 | 31,14,644 |
| 8. | Public Institution | 1,30,000 | 00,88,320 | 1,00,000 | 75,000 |
| 9. | Other Miscellaneous | 26,52,105 | 30,21,184 | 17,31,564 | 29,92,439 |
| | Total Expenditure | 1,82,71,948 | 1,84,31,716 | 2,09,46,339 | 2,39,33,806 |
| | Extraordinary & Debt | 93,361 | 72,161 | 1,33,637 | 98,073 |
| | Balance | 22,27,450 | 17,59,628 | 14,31,337 | 41,86,310 |
| | Grand Total | 2,05,92,759 | 2,02,63,505 | 2,25,11,305 | 2,82,18,189 |

# 2

# Problems and Public Interventions: An Analysis of Residents' Perception

## 2.1 Introduction

While facts and figures continue to be the basis of planning for the future, the participation of people in the planning process is increasingly becoming crucial. This change is also manifested in the process of decentralisation, strengthened by the 73rd and 74th constitutional amendments that address the issues relating to empowerment of the people at the grassroots level. In order to capture this slice of reality– of people's perceptions of problems and public interventions–firstly, a group meeting with the elected representatives of the City Board was held which was followed by the members responding to the questionnaire on civic services and infrastructure. Secondly, similar meetings were organised with a cross-section of people in each of the city wards. The residents who attended these meetings also responded to the questionnaire. Thirdly, questionnaires were canvassed among the members of the Hotel Owners' Association and representatives of government and quasi-government organisations and meetings with them were also organised. This section of the report brings out the results of the analysis of this information.

## 2.2 Residents' Perception of Problems Relating to City Infrastructure

The two top ranking common problems identified by all interest groups are unplanned construction resulting in housing shortage for the local population and traffic/parking problems and the ensuing congestion (Table 2.1). While unemployment is identified both by the residents and their representatives as the third major problem, it does not understandably, find a place in the ranking of other interest groups. Among other problems commonly identified is poor infrastructure, including water and power supply. Lack of education and health facilities also emerge in the list of local population and partly in the list of their representatives.

**Table 2.1:** Prioritisation of Problems by Various Interest Groups

| Problems | Rank Residents | Rank Elected Representatives | Rank Officials from Govt. Organisations | Rank Hotel Owner's Assoc. |
|---|---|---|---|---|
| Unplanned Construction and Housing Shortage | 1 | 1 | 1 | 1 |
| Traffic Congestion Parking | 2 | 2 | 2 | 4 |
| Inadequate Water Supply | 5 | 3 | 3 | 6 |
| Poor Sanitation Facility | 6 | 7 | 4 | 5 |
| Poor Infrastructure | 4 | 5 | 5 | 3 |
| Unemployment | 3 | 4 | – | – |
| Lack of Educational Facilities | 7 | 6 | – | – |
| Lack of Health Facility | 8 | – | 7 | – |
| Deforestation/ Forest Laws | – | – | 6 | 2 |

In this background, a detailed analysis of the residents' perception of problems and possible interventions may be taken up. About 95 per cent of the sample residents found the infrastructure facility in the city inadequate (Table 2.2). This perception differs marginally with the income and educational level of the respondents, wherein a lesser proportion of residents with low income and low levels of education find city infrastructure inadequate than residents with higher levels of income and education.

**Table 2.2:** Distribution of Residents' Response to Adequacy of Infrastructure by Educational and Income Levels

| Education Level of Respondents | Adequate | | Inadequate | | Total Numbers |
|---|---|---|---|---|---|
| | Number | Percent | Number | Percent | |
| Educational level | | | | | |
| Illiterate | 1 | 16.67 | 5 | 83.33 | 6 |
| Class 1–5 | 2 | 11.76 | 15 | 88.24 | 17 |
| Class 6–12 | 7 | 5.47 | 121 | 94.53 | 128 |
| Higher Education | 7 | 4.61 | 145 | 95.39 | 152 |
| All Respondents | 17 | 5.61 | 286 | 94.39 | 303 |
| Income Level | | | | | |
| Low | 8 | 7.92 | 93 | 92.08 | 101 |
| Medium | 6 | 3.90 | 148 | 96.10 | 154 |
| High | 3 | 6.25 | 45 | 93.75 | 48 |
| All Respondents | 17 | 5.61 | 286 | 94.39 | 303 |

This high degree of inadequacy of infrastructure, in general, is more or less uniformly found across all types of services, be it civic facilities such as waste disposal, street lighting or others such as supply of potable water, power or provision of education and health services. In practically all cases, perceived inadequacy is associated with income level and educational attainment of the respondents– wherein respondents with higher income and higher level of educational achievement tend to find infrastructure in the city inadequate in relation to those with lower income and low level of educational achievement. In this background, the analysis of adequacy of various facilities has not been stratified according to the background of the respondents. Similar studies of perception of residents in other cities do not indicate this to be a high magnitude of perceived inadequacy of city infrastructure.[1]

*Water and Power Supply*

*The Jal Sansthan* and the Uttar Pradesh State Electricity Board undertake the supply of water and power. The Municipal Board has little role in this. All sections of the residents perceived the quantity of water supplied as inadequate (Table 2.3). It is not, however, clear as to why the quality of water is perceived to be adverse. Information available from the Pollution Control Board finds that the pollution of water in and around Mussoorie is well within permissible limits with reference to all major parameters. With the city experiencing power cuts, particularly during winter of

**Table 2.3:** Distribution of Residents' Perception of Adequacy of Water and Power Supply

| Service | Adequate/ Not adverse | | Inadequate/ Adverse | |
|---|---|---|---|---|
| | Number | Percent | Number | Percent |
| Water supply | 27 | 8.91 | 276 | 91.09 |
| Quality of Water | 57 | 18.81 | 246 | 81.19 |
| Public Water Facility | 39 | 12.87 | 264 | 87.13 |
| Supply of Electricity | 50 | 16.50 | 253 | 83.50 |
| Service of the Electricity Department | 38 | 12.54 | 265 | 87.46 |
| Manpower in Electricity Department | 250 | 84.18 | 47 | 15.82 |
| Impact of power cut | 265 | 87.46 | 38 | 12.54 |

[1] Ramachandran, H (1985), Living Conditions in Bangalore–An Analysis of Residents' Perception, *Ekistics*.

about 4 to 6 hours per day, it is natural that a large proportion of the residents finds the power supply to be inadequate. The services of the electricity department are also rated poor, while the manpower with the electricity department is perceived to be adequate. It should also be noted that the power cuts do not seem to affect productivity, largely because there is very little industrial activity and much of the city economy is based on the service sector.

### *Civic Services*

Among the civic services, availability of street lighting, rain shelters and the number of *safai karamcharis* are found to be adequate by a majority of the residents (Table 2.4). A degree of class bias both in the provision of some of these services and the expectations of the various strata of society is indicated, wherein about 30 to 50 per cent of the poorer strata find street lighting inadequate, whereas the corresponding proportion from richer segments is between 16 and 21 per cent. The residents, irrespective of class, do not seem to be very dissatisfied with the effectiveness of *safai karamacharis*. It should, however, be realised that unlike large

**Table 2.4:** Distribution of Residents' Response to Adequacy of Civic Services

| Civic Service | Adequate/ Not adverse | | Inadequate/ Adverse | |
|---|---|---|---|---|
| | Number | Percent | Number | Percent |
| Street lighting | 227 | 74.92 | 76 | 25.08 |
| Rain Shelters | 207 | 68.32 | 96 | 31.68 |
| Public Bins | 39 | 12.87 | 264 | 87.13 |
| Safai Karamchari | 207 | 68.32 | 96 | 31.68 |
| Incineration Facility | 50 | 16.50 | 253 | 83.50 |
| Public Urinals | 34 | 11.22 | 269 | 87.78 |
| Cleanliness in Public Urinals | 16 | 5.28 | 287 | 94.72 |
| Stray dogs/monkey menace | 45 | 14.85 | 258 | 85.15 |
| Regulations for Pets | 51 | 16.83 | 252 | 83.17 |
| Veterinary Facilities | 39 | 12.87 | 264 | 87.13 |

cities of the plains, there is very little class-based segregation in the residential land use of Mussoorie. All other civic services, such as availability of public bins, incineration facility, availability of public urinals and their cleanliness, veterinary facilities etc. are perceived to be inadequate by a large proportion of residents.

*Transport and Congestion*

Though transport, parking and traffic-related problems reach their peak during the tourist season, these problems are found to be serious by a large proportion of the residents throughout the year (Table 2.5). A large part of the transport requirement of the residents is in the context of access to places outside Mussoorie and not so much for movement within the city. However, the problems in both contexts are accentuated during the tourist season. As a consequence, a large proportion of the residents, particularly, during the peak tourist season perceives problems such as lack of bus and taxi facility, traffic management, crowding and congestion. The quality of roads and their maintenance is found to be inadequate by over 70 per cent of the residents. At the same time, since a

**Table 2.5:** Distribution of Residents' Response to Adequacy of Transport Services

| Transport Facility | Adequate/ Not adverse | | Inadequate/ Adverse | |
|---|---|---|---|---|
| | Number | Percent | Number | Percent |
| Over-crowding in Bus Stands | 157 | 51.82 | 146 | 48.18 |
| Bus/Taxi Facility | 29 | 9.57 | 274 | 90.43 |
| Traffic Management | 35 | 11.55 | 268 | 88.45 |
| Parking Space | 22 | 7.26 | 281 | 92.74 |
| Crowding in Mall (peak ) | 218 | 71.95 | 85 | 28.05 |
| Road Congestion (peak) | 25 | 8.25 | 278 | 91.75 |
| Road Network | 80 | 26.40 | 223 | 73.60 |
| Road Maintenance | 34 | 11.22 | 269 | 88.78 |

significant proportion of the earnings are from tourists, many of the residents do not perceive crowding in the city commercial area during the peak season, to be a problem. For the same reason crowding in the bus stands is also perceived as adverse only by less than half of the sample respondents. Here again, there are class differences in perception – a larger proportion (53 to 67 per cent) of the poorer sections of the residents (who are dependent on public transport to a larger extent) find bus stands overcrowded as compared to the richer sections (37 per cent to 45 per cent).

### *Education and Health Facilities*

Inadequacy in basic social infrastructure such as educational and health facilities is also perceived by a large proportion of the residents (Table 2.6). More than three fourths of the respondents rate capacity in the city schools as inadequate, in a city that houses excellent schooling facility. That there is a lack of affordable schools is clear, particularly when we consider that less than 30 per cent of the respondents feel that there is overcrowding in schools in the city. Similarly, over 85 per cent of the respondents perceive health facilities, both regular and specialised, are inadequate.

**Table 2.6:** Distribution of Residents' Response to Adequacy of Education and Health Facility

| Social Infrastructure | Adequate/ Not adverse | | Inadequate/ Adverse | |
|---|---|---|---|---|
| | Number | Percent | Number | Percent |
| Intake Capacity of Schools | 71 | 23.43 | 232 | 76.57 |
| Crowding in Schools | 86 | 28.38 | 217 | 71.62 |
| Hospital Facility | 45 | 14.85 | 258 | 85.15 |
| Specialised Medical Facility | 51 | 16.83 | 252 | 83.17 |

### *Land Development and Housing*

One of the most vociferously expressed problems relates to the shortage of housing for the resident population and the nature and implementation of the building by-laws. This has been expressed emphatically in all the participatory meetings held in various city wards as well as with the interest groups and evidenced by the response to the questionnaire (Table 2.7). Almost 90 per cent of the respondents perceive that the role of the development agency, MDDA, has been negative and has encouraged speculative and absentee housing development, pushed up the land values and marginalised the residents' requirements. The impact is also perceived to be adverse on city aesthetics and a contributory factor to congestion. A significant proportion of the residents has also noted the lack of entertainment and sports facilities.

**Table 2.7:** Distribution of Residents' Response to City Development and Housing

| Development Issue | Not adverse | | Adverse | |
|---|---|---|---|---|
| | Number | Percent | Number | Percent |
| Development Planning | 31 | 10.23 | 272 | 89.77 |
| Construction-related Congestion | 64 | 21.12 | 239 | 78.88 |
| Construction-related City Aesthetics | 57 | 18.81 | 246 | 81.19 |
| Implementation of Building by-laws | 93 | 30.69 | 210 | 69.31 |
| Entertainment Facility | 30 | 9.90 | 273 | 90.10 |
| Accessibility to Picnic Spots | 85 | 28.05 | 218 | 71.95 |
| Sports Facility | 16 | 5.28 | 287 | 94.72 |

*Note:* Figures in brackets denote percentages.

## *Pollution and Landslides*

The difference in air pollution during the tourist season and off-season is perceived to be significant. Close to 90 per cent of the respondents believe that the city air is not polluted during the off season as against about 80 per cent of the respondents who believe that the air is polluted during the tourist season (Table 2.8). It should be noted that much of the air pollution in the city is through emission from automobiles that reaches a very high density during the tourist season. This dimension is also associated with noise pollution. Land degradation resulting in landslides—specifically along the most frequently used Dehradun-Mussoorie road, is found to be frequent by about 95 per cent of the respondents.

**Table 2.8:** Distribution of Residents' Perception of Pollution

| Nature of Pollution | Not adverse | | Adverse | |
|---|---|---|---|---|
| | Number | Percent | Number | Percent |
| Air Pollution | 267 | 88.12 | 36 | 11.88 |
| Air Pollution (Peak Season) | 58 | 19.14 | 245 | 80.86 |
| Noise Pollution | 255 | 84.16 | 48 | 15.84 |
| Noise Pollution (Peak Season) | 29 | 9.57 | 274 | 90.43 |
| Landslides (Dehradun-Mussoorie Road) | 11 | 5.05 | 207 | 94.95 |

*Perceived Indicative Solutions*

For the problems prioritised (see Table 2.1), various interest groups also indicated a broad approach to the solution. These are summarised in Table 2.9.

## 2.3 Intra-City Disparity in Infrastructure and Services

Given the variations in the location of different city wards *vis-a-vis* elevation, slope and the city centre, the nature of problems perceived differ from ward to ward. These differences have been documented through ward-wise meetings as well as the analysis of the residents' response to questionnaires that were canvassed. The two top ranking problems identified by residents of the wards as well as what may be termed as 'problem wards' are highlighted here.

Eight of the 25 city wards indicate problems relating to housing and land-development as the first major problem (Figures 2 and 3) of Mussoorie. Such wards are mainly located in what may be termed as the city centre, which appears to be already densely built-up. Four of the city wards, particularly those in the outlying areas rate

**Table 2.9:** Major Indicative Solutions Offered by Respondents

| Problems relating to | Major Indicative Solutions Offered |
|---|---|
| Road Congestion/ Parking | Essentially repair and maintenance, railings, expansion of parking facilities at Masonic Lodge, Library and Kings Craig, additional traffic police during peak season, construction of *Chuna-khala* road, development of 500 acre land purchased by tourism department in *Jharipani,* bus connection between *Jharipani* and Mussoorie be improved. |
| Water Shortage | Expansion of storage and pumping facility, better distribution system within the city |
| Power Shortage | Power cuts in hill areas should be totally stopped, modernisation of Galogi power house, better maintenance of transformers. |
| Waste Management | Better monitoring of *safai karamcharis,* expansion of sewerage system, shifting of dumping grounds away from built-up area, construction of more public toilets and pay toilets, awareness campaigns, provision of proper stable facility for horses/mules. |
| Land Development/ Housing | Removal of encroachments, construction of houses for local population should be permitted, the role of the MDDA should be restricted to outside City Board limits and the development role within the city should be restored to the City Board, ban on commercial construction within and on the periphery of Mussoorie, redefining of forest area within the city limits. |
| Unemployment | Implementation of self-employment schemes, establishing small scale activities such as dairy, poultry, etc., at UPSMDC complex. |
| Health | Augmentation of health facility including special services. |
| Education | Establishment of affordable schools and higher education/vocational education facility. |

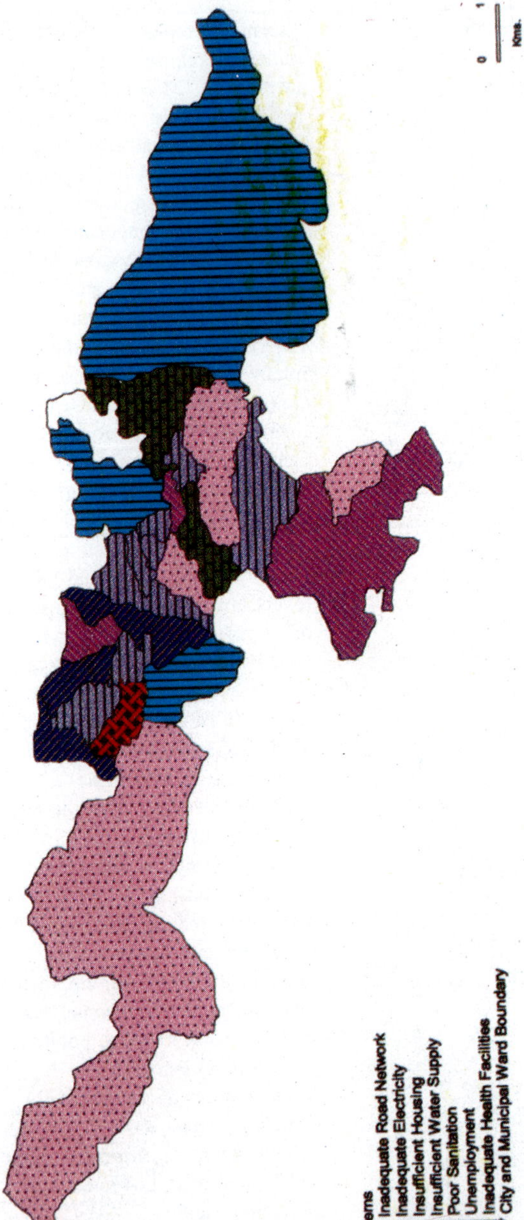


**Fig. 2:** Mussoorie—Ranking of Problems by Residents (First Ranking Problems)

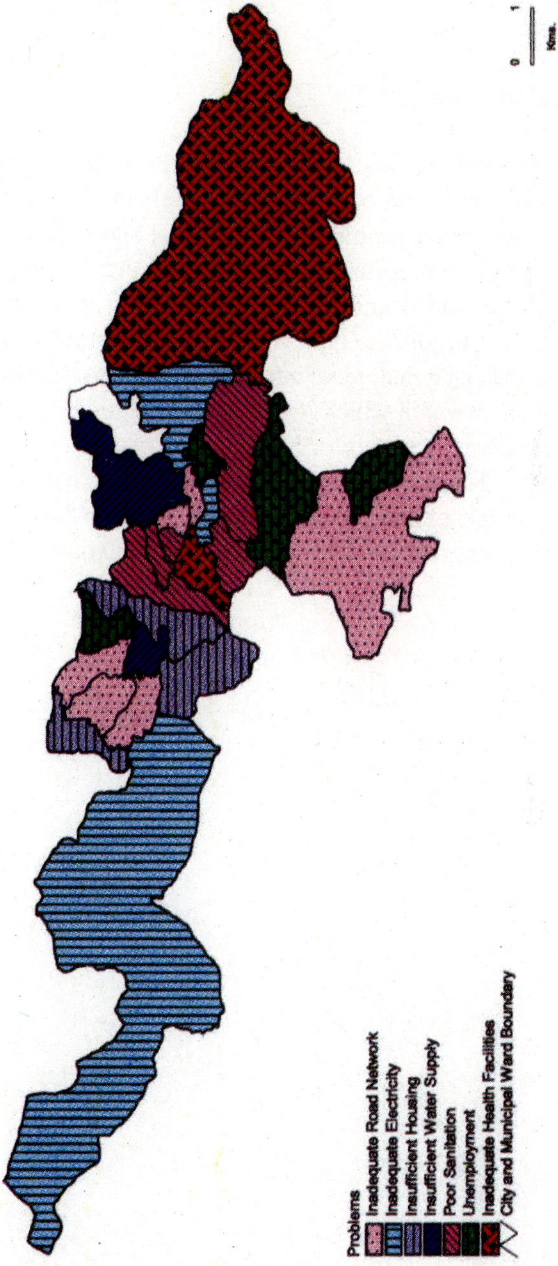


**Fig. 3 :** Mussoorie— Ranking of Problems by Residents (Second Ranking Problems)

the poor and inadequate road network as the priority problem, while insufficient electricity and water supply are perceived in at least three of the 25 wards.

## 2.4 Concluding Remarks

The above analysis basically enumerates the problems of the residents of Mussoorie, as perceived by them. There is no indication of the magnitude of the problem, the degree of shortage in urban services, the level of stress on available infrastructure, nor of the available financial and technical capacity of the city administration to respond to the problems. Nevertheless, this approach gives a sound basis on which quantitative information needs to be generated and indicates the ways in which it needs to be analysed in order to measure the carrying capacity of Mussoorie. The approach would be to develop indicators that reflect the residents' problems. The subsequent sections of this report attempt to measure the stress on available city infrastructure *vis-a-vis* the city population.

# 3

# On Measuring Carrying Capacity of Mussoorie

## 3.1 Estimating the City Population

The carrying capacity of an urban eco-system as a whole or part thereof is, essentially, a function of its population among other things. The first problem involved in estimating the carrying capacity of the Mussoorie U.A. was the paucity of reliable population data. The last census conducted in 1991, as well as the previous one conducted in the winter of 1981, seemed to have considerably underestimated the population as indicated in Chapter I. Census population figures for 1981 and 1991 are given in Table 3.1.

However, a survey conducted by the Mussoorie Water Supply and Sewerage Board in 1983 placed the estimated population at 31,500 with an estimated floating population of 20,200. It appears that while the census figures were underestimates, those of the Water Supply and Sewerage Board were overestimates. This necessitated a different approach to an estimation of the current (1997) population on the basis of which projections could be made. The various segments of the population of the Mussoorie U.A. have been estimated as follows.

**Table 3.1:** Census Population of Municipal Board and Cantonment

| | 1991 | 1981 |
|---|---|---|
| Mussoorie U.A. | 29,629 | 18,233 |
| Landour Cantonment | 2,907 | 1,910 |
| Mussoorie (M.B.) | 26,722 | 16,323 |

*Population of the Municipal Board Area*

The voters list for the City Board election (updated in 1995) was acquired. The number of voters registered was 17,770 i.e. equivalent to the 18 plus population. As age-wise population break-up was not available for the Mussoorie U.A., the age-wise population data for a similar hill urban area (Shimla district, Himachal Pradesh) was therefore, obtained. The proportion of population in the 18 years plus age group was 55 per cent. Taking 17,770 as equivalent to 55 per cent of the population, the total population of area under the jurisdiction of the Mussoorie Municipal Board was estimated at 31,218.

*Population of the Cantonment Area*

Census figures place the population of Landour Cantonment at 2,907 in 1991. The Landour Cantonment Board has confirmed these figures. The Board indicated an annual increase of approximately 5 persons. On this basis, the population was extrapolated as given in Table 3.2. The Landour Cantonment Board further notified us of a decrease of 150 persons with the closing of the PPCL Corporation during 1997. Thus, the 1997 population estimate works out to 2,787.

**Table 3.2 :** Population Estimates for Landour Cantonment Area

| Year | Population |
|---|---|
| 1991 | 2907 |
| 1992 | 2912 |
| 1993 | 2917 |
| 1994 | 2922 |
| 1995 | 2927 |
| 1996 | 2932 |
| 1997 | [2937 (-150)] = 2787 |

*Tibetan Population*

As the Tibetan population of Mussoorie does not figure in the voters' list, the number of Tibetans registered with the Local Intelligence Office, Police Department, Dehradun, was obtained. The Tibetan population above 16 years of age was given as 656. To this figure, we added the number of Tibetan children studying as day scholars in the two schools for Tibetans in Mussoorie i.e. (1) Central School for Tibetans 170 and (2) Tibetan Homes School 33. The total Tibetan population was thus placed at 656 + 203 = 859.

*Boarders in Tibetan Schools*

The total number of boarders in the two Tibetan schools is as follows: (1) Central School for Tibetans had 493 children and (b) Tibetan Homes School had 1,268 children. The total Tibetan population was thus placed at 1,761. Boarders in Tibetan schools were assessed separately as their vacation timings are different from the public schools and would have an impact on the month-wise population estimates made later.

*Boarders in Public Schools*

A survey of all 16 boarding schools in Mussoorie was carried out in order to ascertain the number of children residing in them. The number of boarders in Mussoorie in 1997 was 6,028. As the number of boarding schools in the 1983 survey was also 16, it is assumed that the total number of children in boarding schools would remain around the same figure i.e. 6,028 for the period under study. It is not feasible to project the number of new boarding schools that may be established in the future.

*Total permanent population estimates for 1997*

Final population estimates for 1997 were arrived at in a disaggregated manner—for the winter months and for the rest of the year. As a fairly substantial proportion of the residents of Mussoorie move down to the plains during the winter months i.e., December, January and February, it was felt that separate population estimates should be made for the three winter months and for the remainder of the year. The steps involved in this are:

1. The 1981 Census (which was conducted during winter) places the population of Mussoorie at 18,241. The 1991 Census figures are 29,629 which works out to a growth rate of 6.2 per cent per annum between 1981 and 1991. Based on this growth rate, the Census figures for 1981 were projected to 1983 as follows:

| Population of Mussoorie U.A. | |
|---|---|
| 1981 | 18,241 |
| 1982 | 18,241 + 1,131 = 19,372 |
| 1983 | 19,372 + 1,201 = 20,573 |

2. The projected population figures for 1983 were then compared with the survey figures for 1983 Water Supply and Sewerage Board to estimate the proportion of residents who stay in the U.A. throughout the winter. The survey of 1983 (in season) estimated the population at 31,398 and that projected on the basis of the Census during winter worked out to 20,573. Thus the proportion of winter population to population during the rest of the year is 65 per cent. Thus the total resident population is placed at 23,486 (i.e. 65 per cent of 36,132 in 1997).
3. Following from the above calculations, the population estimates for the winter months (1997) are detailed below: December and February: Resident population (22,662) + Boarders in Tibetan Schools (1,761) = 24,423
   January: Resident population (22,662) + Boarders in SOS children's village (1,268) = 23,930.
4. The estimates of in-season population are as follows:

| | |
|---|---|
| Total residents | 34,864 |
| Boarders in public schools | 4,267 |
| Tibetan boarders | 1,761 |
| Total | 40,892 |

5. The population in winter and the rest of the year (1997) is thus estimated to be:

| | |
|---|---|
| December | 24,423 |
| January | 23,930 |
| February | 24,423 |
| Other Months | 40,892 |

## 3.2 Population Projections

Based on the population estimates for 1997, an attempt has been made to project the population for the next 24 years i.e., to the year 2021. This was done with a view to estimating the carrying capacity of Mussoorie with reference to essential elements of the infrastructure such as water supply, waste disposal, availability of beds for tourists, etc. Population projections have been made at five-year intervals from 1997 to 2021.

*Estimating the population growth rate*

Census data for the population of the Mussoorie U.A. was collected for a period of 90 years i.e., for the census years beginning from 1901 at ten-year intervals to 1991. The figures were plotted on a graph and trend lines were fitted—both linear and logarithmic (Figure 4). The major difficulty in projecting the population of Mussoorie arises from the fact that during the period of nearly a century under consideration there have been several sharp dips or falls in population (Table 3.3), and the decadal growth rate has been fluctuating correspondingly.

Based on the linear trend line—best fitting the population distribution over the period 1901-1991— the expected population for the year 1997 was 24,737. On the basis of our own population estimates for 1997 detailed above, the winter population varies between 24,737 and 25,247 which closely approximates the forecast based on the linear trend line. However, to estimate the growth rate of the population several calculations were made, before adopting what seemed to be the most realistic figure.

**Table 3.3** : Mussoorie Urban Agglomeration—Population Growth (1901-91)

| Name | Population and Growth Rate | | | | | | | | | |
|---|---|---|---|---|---|---|---|---|---|---|
| | 1901 | 1911 | 1921 | 1931 | 1941 | 1951 | 1961 | 1971 | 1981 | 1991 |
| Mussoorie U.A. | 6461 | 8052 | 9702 | 6116 | 7172 | 8550 | 11238 | 20389 | 18233 | 29629 |
| | | (24.62) | (24.49) | (-36.96) | (17.27) | (19.21) | (31.44) | (81.43) | (-10.57) | (62.50) |
| (i) Landour Cantonment | 1720 | 1500 | 1405 | 1150 | 1206 | 1417 | 1389 | 2351 | 1910 | 2907 |
| | | (-12.79) | (-6.33) | (-18.15) | (4.87) | (17.50) | (-1.98) | (69.26) | (-18.76) | (52.20) |
| (ii) Mussoorie M.B. | 4741 | 6552 | 8297 | 4966 | 5966 | 7133 | 9849 | 18038 | 16323 | 26722 |
| | | (38.20) | (26.63) | (-40.15) | (20.14) | (19.56) | (38.08) | (83.15) | (-9.51) | (63.71) |

*Note:* Figures in parentheses indicate per cent decadal growth.

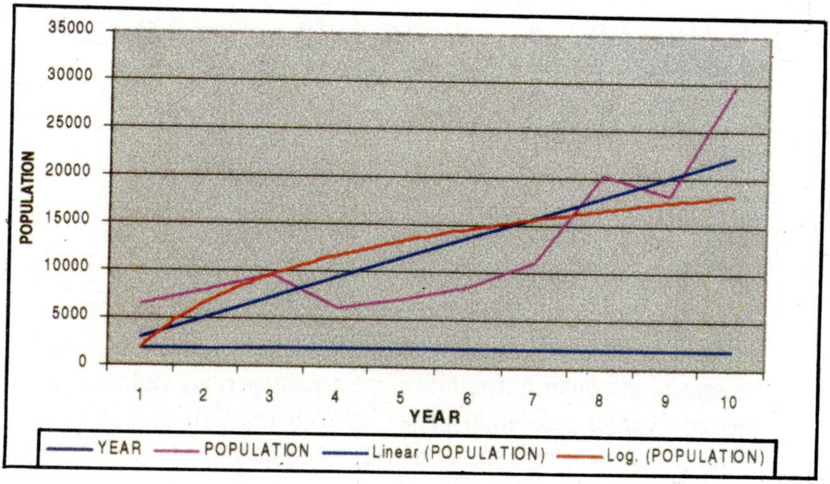


**Fig. 4:** Mussoorie—Population Growth (1901-1991)

**Method I**: The growth rate of Mussoorie U.A. between Census 1981 and Census 1991 was calculated as follows:

$$\frac{\text{Pop.}_{91}\ (29{,}629) - \text{Pop.}_{81}\ (18{,}233)}{\text{Pop.}_{81}\ (18{,}233)} \times 100$$

= 62.50 per cent (or 6.25% per annum)

**Method II:** The growth rate based on the linear trend line is as follows:

| | |
|---|---|
| Population 2001 | = 25,263 |
| Population 1997 | = 24,737 |
| Growth rate for 4 years | = 2.13 per cent |
| Growth rate per year | = 0.53 per cent |

**Method III:** The growth rate based on Survey (1983) and our estimate (1997).

| | |
|---|---|
| Population (1997) | = 34,864 (Mussoorie U.A.) |
| Survey (1983) | = 25,456 (Mussoorie U.A.) |

Growth rate for 14 years = 36.96
Growth rate per year = 2.6 per cent

**Method IV:** The growth rate based on Survey (1983) and 1991 Census

Population (1991) = 29,629
Survey (1983) = 25,456
Growth rate for 8 years = 16.39 per cent
Growth rate per year= 2.05 per cent

Of the four different growth rates, growth rate I i.e. 6.25 per cent is unrealistically high while that based on the linear trend is too low (probably because of the fact that the Census had been conducted in winter during the some census periods and accounts for only winter residents). Thus, growth rates III and IV were selected and two separate scenarios were generated:

**Scenario I:** based on a growth rate of 2.05 per cent per annum.
**Scenario II:** based on a growth rate of 2.60 per cent per annum

Estimating the growth rate of Tourists. In order to project the number of tourists visiting Mussoorie over the next 25 years, data on the number of tourists, both Indian and foreign—entering Mussoorie every month for the years 1990 to 1996 was collected from the Department of Tourism, Government of Uttar Pradesh (see Appendix 4). The growth rate of the number of tourists visiting Mussoorie was then calculated as follows:

$$\frac{\text{Number of tourists (1996)} - \text{Number of tourists (1990)}}{\text{Number of Tourists (1990)}}$$

which gives an annual growth rate of 0.6 per cent.

The number of tourists expected to visit the Mussoorie U.A. at five-yearly intervals was then calculated on an annual growth rate of 0.6 per cent. Thus, the projection of various segments of Mussoorie population is given in Table 3.4.

**Table 3.4** : Population Projections for Mussoorie U.A. 1997-2021

| Year | Permanent Population Scenario I Annual Growth Rate 2.05% | Permanent Population Scenario II Annual Growth Rate : 2.6% | Tourists Annual Growth Rate 0.6 percent | Boarders in Residential Schools |
|---|---|---|---|---|
| 1997 | 34864 | 34864 | 1427514 | 6028 |
| 2001 | 37723 | 38490 | 1461774 | 6028 |
| 2006 | 41590 | 43494 | 1505627 | 6028 |
| 2011 | 45853 | 49148 | 1550796 | 6028 |
| 2016 | 50553 | 55537 | 1597320 | 6028 |
| 2021 | 55735 | 62757 | 1645240 | 6028 |

## 3.3 Estimating the Population Pressure

Once the population projections for the different segments of population i.e. permanent population, boarders in residential schools and tourists were made, the next step involved estimating the population pressure on the town. As Mussoorie, being a tourist town, shows sharp seasonal fluctuations in population, it was decided to estimate population pressure on a monthly basis. This would also aid in estimating requirements of water, waste disposal, hotel beds, etc., on a month to month basis.

For each of the selected years, population projections were made on the basis of the two growth rates selected—2.05 per cent per annum for Scenario I and 2.6 per cent per annum for Scenario II.

### *Permanent Population Estimates*

Based on the projected permanent population figure for any particular year, the total in-season permanent population was derived by adding to the permanent population the total boarders in residential schools. Off-season in Mussoorie coincides with the winter months of December, January and February. The method of estimating the permanent population for these three months is as follows:

**For December and February**

Resident population (65 per cent of the permanent population + all boarders in Tibetan schools), e.g. for January 1997:
Resident population = 65 per cent of 34,864
= 22,662 + 1,761 = 24,423

**For January**

Resident population (65 per cent of the permanent population + boarders in SOS children's village who stay in Mussoorie throughout the year.
Resident population = 65 per cent of the 34,864
= 22,662 + 1,268 = 23,930

*Estimates of Tourist Pressure*

Based on the average month-wise arrival of tourists from 1990 to 1996, the proportion of tourists arriving in any one month of the year to the total tourists arriving annually was calculated (Table 3.5).

The projected figures for the years 1997, 2001, 2006, 2011, 2016 and 2021 were then split proportionately into tourist arrivals per month. As the average tourist spends three days in Mussoorie in May and June and two days in the other months, tourist arrivals per month were converted into the number of tourist days. The final figure of tourist pressure was obtained by dividing the number of tourist days per month by the number of days in the month to obtain the number of tourists per day in that particular month as illustrated below:

e.g., To estimate tourist pressure for January 1997
1. Total tourist arrivals in January = 45,758
2. Tourist days @ 2 per tourist = 91,516
3. Tourists per day = 91,516/31
= 2,952.

*Final Estimates of Population Pressure*

To obtain the final estimates of population pressure per month, the permanent population for each month was added to the tourists, per day in the month (See Appendix 5).

**Table 3.5:** Tourists In-flow to Mussoorie (1990-96)

| Year | Average Indian Tourists (1990-96) | Average Foreign Tourists (1990-96) | Average All Tourists (1990-96) | All Tourists as per cent of previous Column Total |
|---|---|---|---|---|
| January | 360372 | 50 | 51532 | 3.6 |
| February | 456536 | 76 | 65295 | 4.6 |
| March | 638536 | 143 | 91362 | 6.4 |
| April | 863634 | 145 | 123521 | 8.7 |
| May | 977264 | 161 | 163038 | 11.5 |
| June | 1755263 | 140 | 281017 | 19.8 |
| July | 1323500 | 165 | 274081 | 19.3 |
| August | 475000 | 144 | 79310 | 5.6 |
| September | 409000 | 166 | 68332 | 4.8 |
| October | 500000 | 128 | 83461 | 5.9 |
| November | 437000 | 102 | 72935 | 5.1 |
| December | 407000 | 75 | 67908 | 4.8 |

## 3.4 Carrying Capacity with Reference to Basic Needs

This section attempts to estimate the carrying capacity of the Mussoorie U.A. with respect to basic needs such as water, power, waste disposal, availability of hospital beds, housing, hotel beds for tourists, etc. The methodology followed here is simple one — that of projecting month-wise demands for each amenity as a function of the per capita requirement and the population projected for that month. As the current supply of the amenity is known, anticipating shortfalls in supply in the future becomes simple.

**Table 3.6:** Projected Water Demand @ 150 lpcd (Scenario - I)

| Months | 1997 | 2001 | 2006 | 2011 | 2016 | 2021 |
|---|---|---|---|---|---|---|
| January | 4086750 | 4373450 | 4769700 | 5201100 | 5675550 | 6197550 |
| February | 4366950 | 4662600 | 5061150 | 5488800 | 5953050 | 6509250 |
| March | **7017900** | **7018050** | **8075250** | **8742600** | **9476400** | **10283400** |
| April | **7375800** | **7834350** | **8452650** | **9131400** | **9876750** | **10695750** |
| May | **8516850** | **8945700** | **9656100** | **10371000** | **11153700** | **12010950** |
| June | **10373550** | **10904100** | **11614350** | **12388050** | **13231200** | **14150850** |
| July | **8800050** | **9292800** | **9954750** | **10678650** | **11470500** | **12337350** |
| August | 6907350 | 7330500 | 8003700 | 8622600 | 9352800 | 10156050 |
| September | 6819000 | 7264350 | 7864950 | 8526450 | 9253800 | 10054200 |
| October | 6948900 | 7397250 | 8002350 | 8667600 | 9399150 | 10203750 |
| November | 6861750 | 7308150 | 7910550 | 8573100 | 9301800 | 10103550 |
| December | 4326600 | 4621200 | 5018550 | 5455050 | 5934900 | 6462600 |

Current supply is 70,00,000 lpcd.

*Note :* Bold figures indicate demand in excess of current supply.

Current carrying capacity = **46,666.** lpcd

**Table 3.7:** Projected Water Demand @ 150 lpcd (Scenario - II)

| Month | 1997 | 2001 | 2006 | 2011 | 2016 | 2021 |
|---|---|---|---|---|---|---|
| January | 4086750 | 4452150 | 4955400 | 5522400 | 6161550 | 6882150 |
| February | 4366950 | 4737300 | 5246850 | 5810100 | 6439050 | 7193850 |
| March | **7017900** | **7583100** | **8360850** | **9236850** | **1022400** | **11336700** |
| April | **7375800** | **7949400** | **8738250** | **9625650** | **10624350** | **11749050** |
| May | **8516850** | **9060750** | **9941700** | **10865250** | **11901300** | **13064250** |
| June | **10373550** | **11019150** | **11904450** | **12882300** | **13978800** | **15204150** |
| July | **8800050** | **9407850** | **10240350** | **11172900** | **12218100** | **13390650** |
| August | **6907350** | **7445550** | **8244300** | **9116850** | **10100400** | **11209350** |
| September | **6819000** | **7379400** | **8150550** | **9020700** | **10001400** | **11107500** |
| October | **6948900** | **7512300** | **8287950** | **9161900** | **10146750** | **11257050** |
| November | 6861750 | 7423200 | 8196150 | 9067350 | 10049400 | 11156850 |
| December | 4326600 | 4695900 | 5204250 | 5776350 | 6420900 | 7147200 |

Current Supply is 70,00,000 lpcd.

*Note*: Bold figures indicate demand in excess of supply.

Current carrying capacity = **46,666.** lpcd

### *Projected Demand for Water*

Water requirements for the U.A. have been worked out on the basis of 150 lpcd (litres per capita per day as given by the office of the Executive Engineer, Uttar Pradesh *Jal Nigam,* Dehradun). Tables 3.6 and 3.7 give the month-wise break-up of water requirement under the two scenarios. The total available supply as of 1997 is 7 million litres per day. As can be seen from Table 3.6, even at present (1997) the available water supply is sufficient only for the three winter months of December, January and February. By June the demand for water is already 1.5 times the available supply. Under Scenario I (Table 3.6), the situation with respect to monthly demand for water remains the same with sufficient water supply being available from December to February even in the year 2021. However, the peak demand for water in the month of June i.e., 141.5 million litres is more than twice the city's carrying capacity. In fact, for all nine months of the year from March to November water demand exceeds the city's supply by about 1.5 times.

Coming to Scenario II (Table 3.7), it is clear that while the existing water supply remains adequate for the period December to February till 2016, by the year 2021, water demand in December and February exceeds the available supply and the supply in January is just barely adequate to meet the demand generated. The peak demand in June will be about twice the current supply. For details of the water supply system in Mussoorie, see Chapter 5.

### *Current Carrying Capacity with Reference to Water Supply*

The current supply of water in Mussoorie is about 70 lakh litres per day. Based on a daily per capita requirement of 150 lpcd, the carrying capacity of the Mussoorie U.A. is estimated as below:

| | |
|---|---|
| Available water supply: | 70,00,000 litres per day |
| Demand per capita: | 150 litres per day |
| Carrying Capacity: | 46,666 persons. |

**Table 3.8:** Estimates of Waste Generation in Mussoorie (Scenario I)

| Months | 1997 | | | 2001 | | | 2006 | | |
|---|---|---|---|---|---|---|---|---|---|
| | Permanent Population @ 350 gm/ cap. (kg.) | Tourists per day @ 400 gm/cap (kg) | Total | Permanent Population @ 350 gm/ cap. (kg.) | Tourists per day @ 400 gm/cap (kg) | Total | Permanent Population @ 350 gm/ cap. (kg.) | Tourists per day @ 400 gm/cap (kg) | Total |
| January | 8375 | 1326 | 9701 | 9026 | 1358 | 10384 | 9905 | 1399 | 1304 |
| February | 8548 | 1876 | 10424 | 9198 | 1921 | 11119 | 10078 | 1979 | 12057 |
| March | 14312 | 1958 | 16270 | 15313 | 2414 | 17727 | 16666 | 2487 | 19153 |
| April | 14312 | 3312 | 17624 | 15313 | 3391 | 18704 | 16666 | 3493 | 20159 |
| May | 14312 | 6355 | 20667 | 15313 | 6355 | 21668 | 16666 | 6702 | 23368 |
| June | 14312 | 11306 | 25618 | 15313 | 11577 | 26890 | 16666 | 11924 | 28590 |
| July | 14312 | 7110 | 21422 | 15313 | 7280 | 22593 | 16666 | 7499 | 24165 |
| August | 14312 | 2063 | 16375 | 15313 | 2028 | 17341 | 16666 | 2176 | 18842 |
| September | 14312 | 1827 | 16139 | 15313 | 1871 | 17184 | 16666 | 1926 | 18592 |
| October | 14312 | 2174 | 16486 | 15313 | 2226 | 17539 | 16666 | 2292 | 18958 |
| November | 14312 | 1941 | 16253 | 15313 | 1988 | 17301 | 16666 | 2048 | 18714 |
| December | 8548 | 1768 | 10316 | 9198 | 1810 | 11008 | 10078 | 1865 | 11943 |

@ 400 gms/capita/day (for tourists) and 350 gms/capita/day (for permanent population).

**Table 3.8:** *(Contd)*.....

| Months | 2011 | | | 2016 | | | 2021 | | |
|---|---|---|---|---|---|---|---|---|---|
| | Permanent Population @ 350 gm/ cap. (kg.) | Tourists per day @ 400 gm/cap (kg) | Total | Permanent Population @ 350 gm/ cap. (kg.) | Tourists per day @ 400 gm/cap (kg) | Total | Permanent Population @ 350 gm/ cap. (kg.) | Tourists per day @ 400 gm/cap (kg) | Total |
| January | 10875 | 1441 | 12316 | 11944 | 1484 | 13428 | 13124 | 1528 | 14652 |
| February | 11048 | 2011 | 13059 | 12117 | 2027 | 14144 | 13296 | 2162 | 15458 |
| March | 18158 | 2561 | 20719 | 19803 | 2638 | 22441 | 21617 | 2717 | 24334 |
| April | 18158 | 3598 | 21756 | 19803 | 3706 | 23509 | 21617 | 3817 | 25434 |
| May | 18158 | 6904 | 25062 | 19803 | 7111 | 26914 | 21617 | 7324 | 28941 |
| June | 18158 | 12282 | 30440 | 19803 | 12651 | 32454 | 21617 | 13030 | 34647 |
| July | 18158 | 7724 | 25882 | 19803 | 7956 | 27759 | 21617 | 8194 | 29811 |
| August | 18158 | 2241 | 20399 | 19803 | 2308 | 22111 | 21617 | 2378 | 23995 |
| September | 18158 | 1985 | 20143 | 19803 | 2044 | 21847 | 21617 | 2106 | 23723 |
| October | 18158 | 2361 | 20519 | 19803 | 2432 | 22235 | 21617 | 2505 | 24122 |
| November | 18158 | 2109 | 20267 | 19803 | 2172 | 21975 | 21617 | 2238 | 23855 |
| December | 11048 | 1921 | 12969 | 12117 | 1978 | 14095 | 13296 | 2038 | 15334 |

**Table 3.9:** Estimates of Waste Generation in Mussoorie (Scenario II)

| Months | 1997 | | | 2001 | | | 2006 | | |
|---|---|---|---|---|---|---|---|---|---|
| | Permanent Population @ 350 gm/ cap. (kg.) | Tourists per day @ 400 gm/cap (kg) | Total | Permanent Population @ 350 gm/ cap. (kg.) | Tourists per day @ 400 gm/cap (kg) | Total | Permanent Population @ 350 gm/ cap. (kg.) | Tourists per day @ 400 gm/cap (kg) | Total |
| January | 8375 | 1326 | 9701 | 9200 | 1358 | 10558 | 10339 | 1399 | 11738 |
| February | 8548 | 1876 | 10424 | 9373 | 1921 | 11294 | 10511 | 1979 | 12490 |
| March | 14312 | 1958 | 16270 | 15581 | 2414 | 17995 | 17333 | 2487 | 19820 |
| April | 14312 | 3312 | 17624 | 15581 | 3391 | 18972 | 17333 | 3493 | 20826 |
| May | 14312 | 6355 | 20667 | 15581 | 6355 | 21936 | 17333 | 6702 | 24035 |
| June | 14312 | 11306 | 25618 | 15581 | 11577 | 27158 | 17333 | 11924 | 29257 |
| July | 14312 | 7110 | 21422 | 15581 | 7280 | 22861 | 17333 | 7499 | 24832 |
| August | 14312 | 2063 | 16375 | 15581 | 2048 | 17629 | 17333 | 2176 | 19509 |
| September | 14312 | 1827 | 16139 | 15581 | 1871 | 17452 | 17333 | 1926 | 19259 |
| October | 14312 | 2174 | 16486 | 15581 | 2226 | 17801 | 17333 | 2292 | 19625 |
| November | 14312 | 1941 | 16253 | 15581 | 1988 | 17569 | 17333 | 2048 | 19381 |
| December | 8548 | 1768 | 10316 | 9373 | 1811 | 11184 | 10511 | 1865 | 12376 |

@ 400 gms/capita/day (for tourists) and 350 gms/capita/day (for permanent population).

**Table 3.9:** *(Contd)..*

| Months | 2011 | | | 2016 | | | 2021 | | |
|---|---|---|---|---|---|---|---|---|---|
| | Permanent Population @ 350 gm/ cap. (kg.) | Tourists per day @ 400 gm/cap (kg) | Total | Permanent Population @ 350 gm/ cap. (kg.) | Tourists per day @ 400 gm/cap (kg) | Total | Permanent Population @ 350 gm/ cap. (kg.) | Tourists per day @ 400 gm/cap (kg) | Total |
| January | 11625 | 1441 | 13066 | 13078 | 1484 | 14562 | 14721 | 1528 | 16249 |
| February | 11797 | 2011 | 13808 | 13251 | 2027 | 15278 | 14894 | 2162 | 17056 |
| March | 19312 | 2561 | 21873 | 21548 | 2638 | 24186 | 24075 | 2717 | 26792 |
| April | 19312 | 3598 | 22910 | 21548 | 3706 | 25254 | 24075 | 3817 | 27892 |
| May | 19312 | 6904 | 26216 | 21548 | 7111 | 28659 | 24075 | 7324 | 31399 |
| June | 19312 | 12282 | 31594 | 21548 | 12651 | 34199 | 24075 | 13030 | 37105 |
| July | 19312 | 7724 | 27036 | 21548 | 7956 | 29504 | 24075 | 8194 | 32269 |
| August | 19312 | 2241 | 21553 | 21548 | 2308 | 23856 | 24075 | 2378 | 26453 |
| September | 19312 | 1985 | 21297 | 21548 | 2044 | 23592 | 24075 | 2106 | 26181 |
| October | 19312 | 2361 | 21673 | 21548 | 2432 | 23980 | 24075 | 2505 | 26580 |
| November | 19312 | 2109 | 21421 | 21548 | 2172 | 23720 | 24075 | 2238 | 26313 |
| December | 11797 | 1921 | 13718 | 13251 | 1978 | 15229 | 14894 | 2038 | 16932 |

### *Projected Needs for Waste Disposal*

The waste generated in the Mussoorie U.A. was estimated separately for the local population and the tourists. The rates selected were 350 gms. of waste per capita per day for residents and 400 gms. per capita per day for tourists. These estimates were based on trials conducted in Mussoorie during May, 1997. (For details see Annexure 2.)

Tables 3.8 and 3.9 show the month-wise generation of waste by the resident population and by the tourists visiting the U.A. In 1997 itself, the Municipality had to cope with a daily waste disposal ranging between 9.7 tonnes in January to 26 tonnes in the peak season (June). By the year 2021, this daily disposal load will be minimum 15 tonnes, and in the peak season it will touch nearly 35 tonnes under Scenario I. If higher growth rates are assumed (Scenario II), the lowest waste load would be 16 tonnes and the highest over 37 tonnes. The city utilises a single dumping site which could become a health risk, with the chances of polluting the Bhilaru Stream which is an important water source for the area.

### *Projected Demand for Beds in Hotels*

As the economy of the Mussoorie U.A. is almost wholly dependent on the tourist trade, it becomes crucial to assess the availability of residential accommodation for tourists. This assessment is normally made on the basis of number of beds available. The Department of Tourism, Mussoorie provided figures on the of beds available in the 140 registered hotels, both government and private, as 8,028. However, as a large number of tourists visiting Mussoorie are from the lower middle and middle income groups, it is necessary to include other places which offer accommodation on a cheaper basis such as *Gurudwaras* and *Dharamshalas*. Accordingly, all eight Dharamshalas listed in the tourist information brochure were visited and information collected on the accommodation available. It is estimated that 1,598 beds are available in these institutions for tourists. All the institutions verified that available beds are full during the season i.e. May and June.

A third group providing hospitality to tourists in Mussoorie is the holiday homes of various public sector undertakings and private firms, and paying guesthouses run by private individuals. A survey of these revealed an additional availability of 218 beds. The total estimated availability of beds for tourists in Mussoorie is given in Table 3.10.

**Table 3.10:** Beds Available for Tourists in Mussoorie

| | |
|---|---|
| 1. Beds in 140 registered hotels | 8,028 |
| 2. Beds in 8 Gurudwaras, Dharamshalas, Musafirkhanas and Temples | 1,598 |
| 3. Beds in holiday homes and paying guest houses | 218 |
| Total Beds Available | **9,844** |

Table 3.11 shows the daily arrival of tourists in Mussoorie in each month. It is obvious that during the season, i.e., in the months of April, May and June the number of incoming tourists is much greater than the accommodation available in the city. Thus, while in May and July, only 62 per cent and 55 per cent of the arriving

**Table 3.11:** Tourist Flow Per Day - Estimates (Mussoorie)

| Month | 1997 | 2001 | 2006 | 2011 | 2016 | 2021 |
|---|---|---|---|---|---|---|
| January | 3315 | 3395 | 3497 | 3602 | 3710 | 3821 |
| February | 4690 | 4803 | 4947 | 5027 | 5067 | 5406 |
| March | 5894 | 6036 | 6217 | 6403 | 6595 | 6793 |
| April | 8280 | 8478 | 8733 | 8995 | 9264 | 9542 |
| **May** | **15887** | **15887** | **16756** | **17259** | **17777** | **18310** |
| **June** | **28265** | **28943** | **29811** | **30706** | **31627** | **32576** |
| **July** | **17775** | **18201** | **18747** | **19310** | **19889** | **20486** |
| August | 5157 | 5119 | 5440 | 5603 | 5771 | 5944 |
| September | 4568 | 4678 | 4815 | 4962 | 5111 | 5265 |
| October | 5434 | 5564 | 5731 | 5903 | 6080 | 6262 |
| November | 4853 | 4970 | 5119 | 5273 | 5431 | 5594 |
| December | 4421 | 4527 | 4663 | 4802 | 4946 | 5095 |

*Note:* Figures in bold indicate that the number of tourists exceeds the available accommodation. Current Carrying Capacity for overnight tourists is 9,844.

tourists can be accommodated, in June, 65 per cent have to return the same day because of lack of accommodation. Thus, while the city bears the burden of providing water and other facilities to these incoming tourists, and faces the regular traffic jams, resulting emission of pollutants, and the responsibility of disposing of the waste generated, it still loses out on the income which would accrue to the town if the tourists spent the night. This would also, perhaps, reduce the constant traffic jams as the day tourists rush to get back to Dehradun before nightfall. On the other hand, there is a considerable excess capacity during several months in the year.

By the year 2016, this shortfall in tourist accommodation would reach a level where only 31 per cent of the incoming tourists can be accommodated in the peak month of June. By 2021, 70 per cent of the tourists arriving in Mussoorie in June would be forced to return to Dehradun or to proceed beyond Mussoorie in search of accommodation. The vast potential in the tourist trade opens up several possibilities—that of increasing the capacity of existing hotels, granting licenses to new hotels which has to be viewed in the context of excess capacity during off-season, or boosting the development of alternative locations like Dhanaulti or Chakrata. This would also necessitate augmenting access to such alternative centres.

### *Current carrying capacity with reference to tourism*

As the total number of beds for tourists in all hotels, guest houses and *dharamshalas* in the town amounts to 9,844, the current carrying capacity of the town with reference to overnight tourists is 9,844.

### *Projected Demand for Hospital Beds*

Town planning norms in India place the requirement of hospitals at one general hospital per 50,000 population. As Mussoorie has three hospitals[1] with capacities ranging from 30 to 53 beds, it was

[1] The Northern Railway Hospital at Jhari Pani has a 58-bed facility. However, it caters to railway personnel only and has thus been excluded from the carrying capacity estimates.

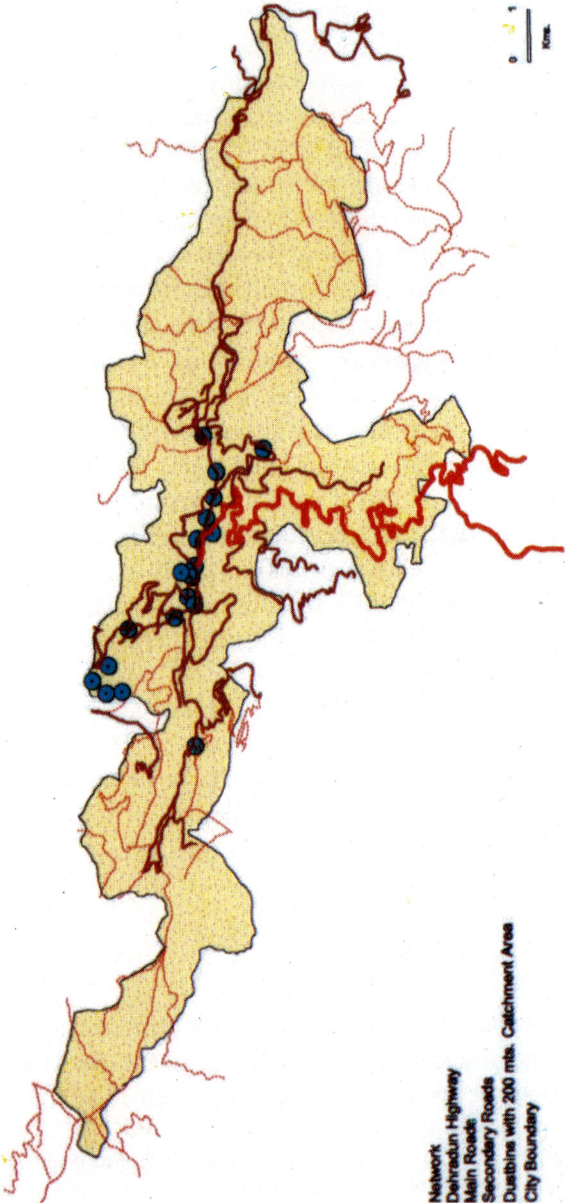


**Fig. 5:** Mussoorie—Spatial Distribution of Dustbins

decided that a more appropriate measure would be the availability of hospital beds against basic requirements. In order to estimate the requirement for hospital beds, one of the largest general hospitals in Dehradun i.e. the Doon Hospital with a capacity of 232 beds was selected. At the rate of one general hospital for 50,000 persons specified by the MDDA as well as Indian town planning norms, the requirement for hospital beds works out to one bed for a population of 215 persons. Thus it was decided to adopt the norm of one hospital bed per 200 persons, as the basis of working out the carrying capacity of Mussoorie with reference to health facilities. The number of hospital beds available in Mussoorie in 1997 is detailed in Table 3.12.

**Table 3.12:** Availability of Hospital Beds in Mussoorie (1997)

| Sl.No. | Name of Hospital | Number of Beds |
|---|---|---|
| 1. | St. Mary's Hospital, Mussoorie | 53 |
| 2. | Civil Hospital, Mussoorie | 30 |
| 3. | Community Hospital, Landour | 50 |
| 4. | Private Nursing Homes | 20 |
| | **Total (Beds)** | **153** |

Projected requirements for hospital beds even under Scenario I (Table 3.13) reveal a shortage of beds from March to November. In fact, even as per 1997 estimates the shortfall in hospital beds in the peak month of June is as much as 193. The pattern remains the same over the next 20 years, though the shortfall in the peak season increases sharply to 319 beds implying that the peak demand for hospital beds is over 200 per cent of the supply. By the year 2021, the requirement for hospital beds exceeds the availability throughout the year.

Under Scenario II (Table 3.14) with the higher population growth rate of 2.6 per cent, the shortfall in hospital bed begins as early as 2001 with sufficient beds being available only in the month

**Table 3.13:** Requirement of Hospital Beds (Scenario I)[2]

| | 1997 | 2001 | 2006 | 2011 | 2016 | 2021 |
|---|---|---|---|---|---|---|
| January | 136 | 146 | **159** | **173** | **189** | **207** |
| February | 146 | **155** | **169** | **183** | **198** | **217** |
| March | **234** | **234** | **269** | **291** | **316** | **343** |
| April | **246** | **261** | **282** | **304** | **329** | **356** |
| May | **284** | **298** | **322** | **346** | **372** | 400 |
| June | **346** | **363** | **287** | **413** | **441** | **472** |
| July | **293** | **310** | **332** | **356** | **382** | 411 |
| August | **230** | **244** | **267** | **287** | **312** | **338** |
| September | **227** | **242** | **262** | **284** | **308** | 335 |
| October | **232** | **247** | **267** | **289** | **313** | **340** |
| November | **229** | **244** | **264** | **286** | **310** | **337** |
| December | 144 | **154** | **167** | **182** | **198** | **215** |

*Note:* Figures in bold indicate that the estimated demand for hospital beds exceeds the current supply. Availability of Hospital Beds - 153. Current Carrying Capacity—30,600 persons.

**Table 3.14:** Requirement of Hospital Beds (Scenario II)

| Months | 1997 | 2001 | 2006 | 2011 | 2016 | 2021 |
|---|---|---|---|---|---|---|
| January | 136 | 148 | **165** | **184** | **205** | 229 |
| February | 146 | **158** | **175** | **194** | **215** | **240** |
| March | **234** | **253** | **279** | **308** | **341** | 378 |
| April | **246** | **265** | **291** | **321** | **354** | **392** |
| May | **284** | **302** | **331** | **362** | **397** | 435 |
| June | **346** | **367** | **397** | **429** | **466** | **507** |
| July | **293** | **314** | **341** | **372** | **407** | 446 |
| August | **230** | **248** | **275** | **304** | **337** | **374** |
| September | **227** | **246** | **272** | **301** | **333** | 370 |
| October | **232** | **250** | **276** | **305** | **338** | **375** |
| November | **229** | **247** | **273** | **302** | **335** | 372 |
| December | 144 | **156** | **173** | **192** | **214** | **238** |

*Note:* Figures in bold indicate that the estimated demand for hospital beds exceeds the current supply. Availability of Hospital Beds - 153. Current Carrying Capacity-30,600 persons.

---

[2] While the large boarding schools do have infirmary beds, these cannot be equated to specialised hospital beds, but must be treated at par with home nursing facilities.

of January. From the year 2006, the number of beds available is insufficient throughout the year. By 2021, the peak requirement in the months of May, June and July is such that available supply meets only 32 per cent to 38 per cent of the requirement.

## Power-Capacity and Consumption

The power supply to Mussoorie is basically via *Purkul Gaon* power station through two feeder lines. The transformers in the power stations have a capacity to supply 14 MVA. In addition, in case of grid failure, there is a standby captive hydel plant at Galogi with a capacity to generate and supply 600 kVA. The maximum load during the peak season is 2,591 KW and during the off-season it is 2,058 KW. The corresponding average loads are 1,372 KW and 1,067 KW.

The average consumption during 1991 was 1.94 KW/hr. and by 1997 it rose to 3.47 KW/hr. The growth in electrical connections and power consumption during the 1990's in the city is given in Table 3.15. The average per consumer load is estimated to be 1.01 KW.

**Table 3.15:** Growth in Number of Electricity Connection and Consumption in Mussoorie

| Year | Number of Domestic Connections | Number of Commercial Connections | Consumption (in million units per month) |
|---|---|---|---|
| 1991 | 2983 | 1304 | 1.4 |
| 1992 | 3200 | 1369 | 1.5 |
| 1993 | 3294 | 1377 | 1.7 |
| 1994 | 3392 | 1401 | 1.9 |
| 1995 | 3522 | 1419 | 2.0 |
| 1996 | 3662 | 1442 | 2.2 |
| 1997 | 3912 | 1483 | 2.5 |

# 4

# The Limits to City Expansion: The Use of Land in the Mussoorie Hills

## 4.1 Introduction

As has been noted earlier the Mussoorie U.A. is located on a ridge and the terrain is extremely rugged, as a result of which the land available for urban development is limited. The U.A. spans an area of about 65.81 square kilometres.[1] In this chapter we analyse: (a) the area under various slope categories, (b) the extent of land under broad land use categories, (c) the distribution of use of land under various categories of slope and (d) change in the pattern of land use between 1990 and 1997.

Cartographic data used in this analysis include:

- Municipal Ward Boundary Map
- Topographical sheets No. 53 F/15 and 53 J/3
- IRS-1B LISS 2 and IRS 1C LISS 3 False Colour Composites of April 1990 and February 1997 with 36.25 meters and 23.5 meters resolution respectively for land use trend
- Aerial Photographs from Survey of India for the year 1984-85 in the scale of 1:12000
- City Map in the scale of 1 cm : 200 meters for the year 1997 provided by Town Planning Department

[1] It may be noted that the extent of area (derived here on the basis of imagery) is short by about 2 square kilometers when compared to the area of Mussoorie given in the Census.

- Guide Maps, Tourist Maps for other general information and comparisons

For preparation of various maps, source maps were digitised using GIS ARC/Info. Land use analysis was undertaken using image-processing software ERDAS-Imagine, Slope Analysis etc. was done with ARC/Info, ARCVIEW-Spatial Analyst and TIN modules.

For the land use map, IRS FCC was rectified using topographical sheets and classified using both Unsupervised and Supervised techniques for optimum results, and outputs were verified with IIRS, Dehradun Forestry Branch and local field information. For various land use categories, tree vegetation constituting more than 40 per cent crown density was categorised as dense forests, 20 per cent to 30 per cent was considered as medium/open forests and below 20 per cent crown density was categorised as scrub/shrub land. Other categories of land use are built-up area, agricultural land and barren land. Since Mussoorie does not have any major perennial streams, water bodies were not introduced as a separate category but are merged with barren land. Land-use maps for 1990 and 1997 are shown in Figs. 6A and 6B.

For calculating slope, a grid of 500m by 500m was superimposed on a contour map with 100m interval. This was digitised and the Spatial Analyst Module was used to calculate and map the degree of slope in each grid. The method identifies maximum rate of change in the altitude, derives slope from the surface and measures degree (or per cent) of slope.

Classified data for both the years were transferred on to a ward-wise base map in order to derive the land use pattern of each city-ward. The slope map was draped over the land use map to understand the relationship between land use and slope. This analysis was done for both 1990 and 1997.

## 4.2 Distribution of Slope

Of the total area of about 66 square kilometres, of the Mussoorie U.A. more than three-fourths has a slope of 30 degrees or more (Table 4.1). It is necessary to emphasise that 100 meter contour

**Fig. 6A** : Mussoorie—Landuse (1990)

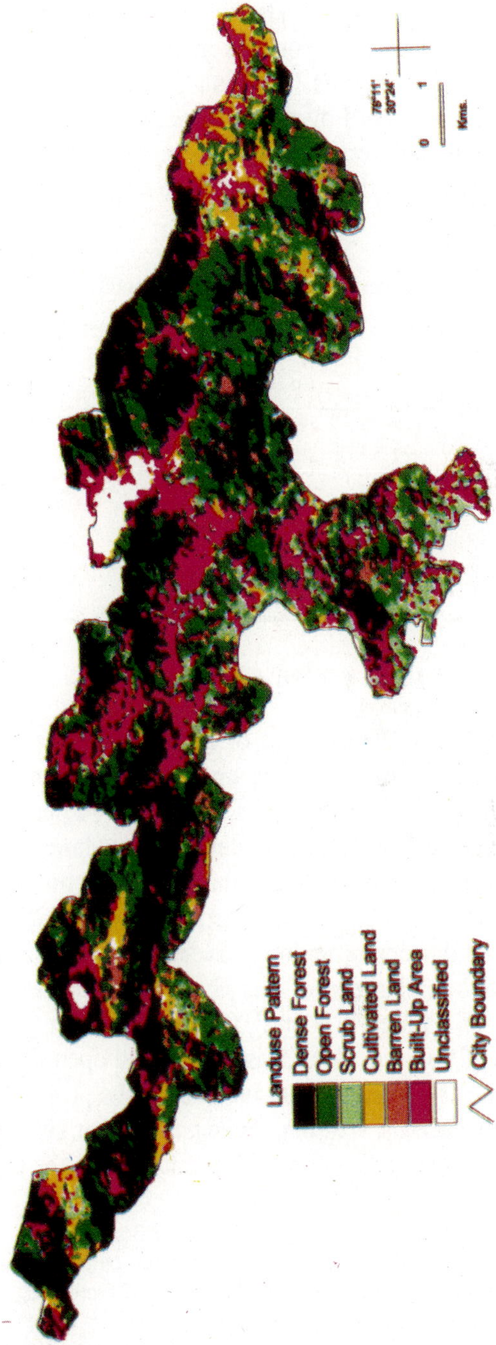


**Fig. 6B :** Mussoorie—Landuse (1997)

interval is not suitable for an urban study. However, the broad lay of the land and area that should not be converted to urban development does emerge from the analysis. Only about 10 square kilometres of the city has less than 20 degrees of slope. Large contiguous areas record over 50 degrees of slopes (see Figure 7). Gently sloping land occurs in smaller parcels. In the south, north and west of the city edge and beyond the city limits we find gently sloping land.

**Table 4.1:** Area and Percentage of Land in Mussoorie by Slope Categories

| Degree of Slope | Area (in Sq. kms) | Percent Area to Total Area |
|---|---|---|
| Below 10 | 5.54 | 8.42 |
| 10-20 | 3.65 | 5.55 |
| 20-30 | 4.23 | 6.43 |
| 30-50 | 10.43 | 15.84 |
| 50-90 | 41.96 | 63.76 |
| Total | 65.81 | 100.00 |

## 4.3 The use of Land ( 1990 and 1997)

A little more than 79 per cent of the land was under forests of various kinds—dense, medium, open and scrub in 1990 (Table 4.2). This proportion has come down to about 72 per cent by 1997. About 6 square kilometres of dense forest-land have become medium/open forest land during this period. Scrub-land has also reduced by about 5 square kilometres. Built-up area (including roads) and land under agriculture have increased during this period.

A visual comparison of this changing land use can be made by a study of Figures 8, 9 and 10 which map out the changes in proportion of built-up area, dense forest and open forest. Most of the city wards show significant increase in the proportion of built-up area. The built-up area that came up between 1990 and 1997 is shown as an inset in Figure 8.[5] The reduction in the proportion of

[5] Since the 1990 imagery had a resolution of 36.25 meters, the land use classification of 1997 imagery took into account only those features which could have been interpreted at a resolution of 36.25 meters.

area under dense forest is however, confined to fewer city wards. Significant reduction is noticeable in Bhadraj, Jabharkhet Jhalki, Jhari Pani, Barlowganj, Hampton Court, Landour South and Nalapani Dhobhighat. It is also worth noting that a few wards have also shown an increase in the proportion of dense forest (Fig.9) such as Bhilaru, Library, Waverley Convent and Allen Wynburg, owing perhaps to plantation activity. In the case of medium and open forest there are city wards which show substantial increase owing largely to a declassification from dense to open forest e.g. Jabharkhet Jhalki, Nalapani Dhobhighat, Polo-ground and Happy valley. On the other hand, there has also been a reduction in the open forest in wards such as Landour North, Beloni hill and Kutchery.

## 4.4 Land use in Relation to Slope

Of the total area under study, over 79 per cent has more than 30 degree slope and, therefore, can be used for urban development only at great risk (Table 4.1). It is clear from the table that if we take only that area which has less than 30 degree slope for construction activity, then we have only 13.42 square kilometres that is available for conversion to built-up area. Table 4.2 shows that the city already records about 11 square kilometres as built-up area. This does not, however, mean that some 2,400 hectares of land is available for further development.

The land with less than 30 degree slope is distributed in several pockets and also contains forest area. Out of the 13.42 square kilometres, as much as 8.73 square kilometres were forests — dense, medium and open — in 1997 (Table 4.3). What is also important to note is that an area of 8.04 square kilometres in areas with a slope of over 30 degrees were already built-up by 1997 (Figure 8). Corresponding built-up area in 1990 was only 6.74 square kilometres. It is possible, however, to argue that the slope analysis is based on a scale larger than that suitable for the purpose of urban development.

A significant proportion of land under dense forest in 1990 has changed into medium or open forest by 1997 (Table 4.3) as indicated by a negative value of 6 percentage points in the former

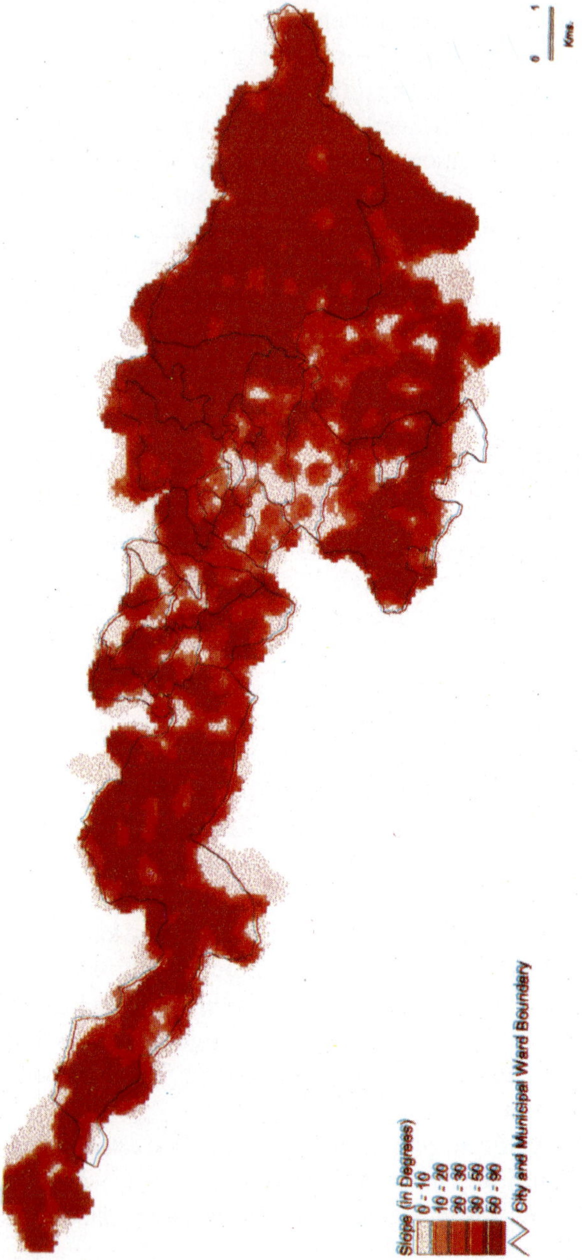


**Fig. 7** : Mussoorie—Slope Analysis

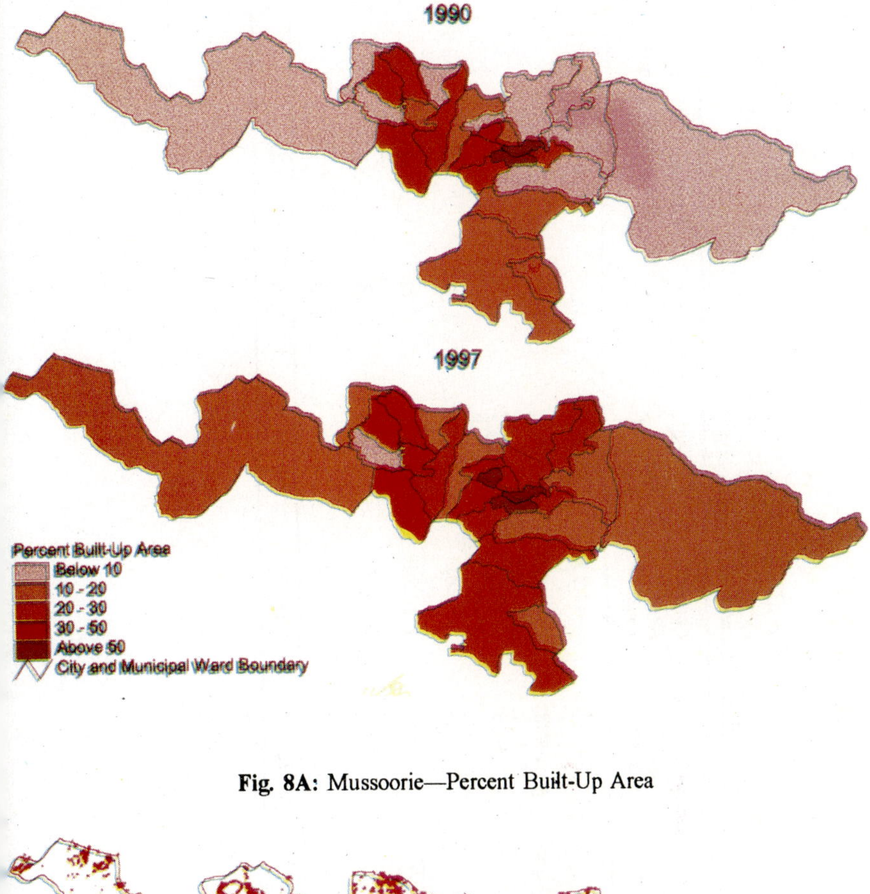


**Fig. 8A:** Mussoorie—Percent Built-Up Area

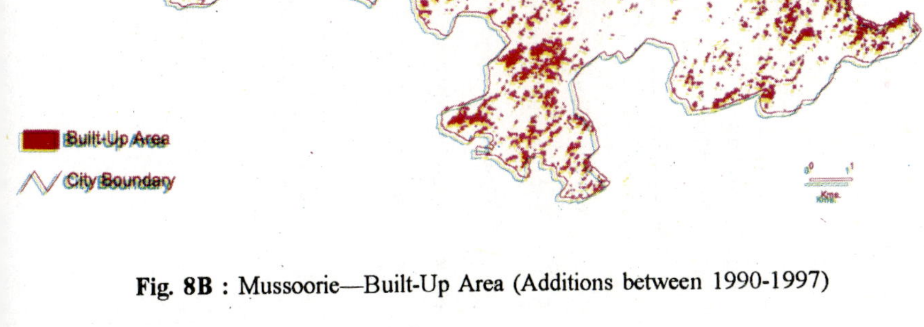


**Fig. 8B :** Mussoorie—Built-Up Area (Additions between 1990-1997)

**Table 4.2:** Change in Landuse in Mussoorie (1990-97)

| Land use Categories | Area (sq.km.) 1990 | Area (sq. km.) 1997 | Percent Area 1990 | Percent Area 1997 | Change in Sq. Kms. | Change Percentage Points |
|---|---|---|---|---|---|---|
| Dense Forest | 29.32 | 23.31 | 44.55 | 35.42 | -6.00 | -9.12 |
| Medium Forest | 11.08 | 18.28 | 16.84 | 27.78 | 7.20 | 10.94 |
| Scrub/Shrub Land | 11.62 | 6.05 | 17.66 | 9.20 | -5.57 | -8.46 |
| Agricultural Land | 2.51 | 3.57 | 3.81 | 5.43 | 1.07 | 1.62 |
| Built-up Area | 6.74 | 11.06 | 10.24 | 16.81 | 4.32 | 6.57 |
| Barren Land | 4.55 | 2.41 | 6.90 | 3.66 | -2.14 | -3.24 |
| Unclassified | 0.00 | 1.10 | 0.00 | 1.70 | 1.10 | 1.70 |
| Total | 65.82 | 65.78 | 100.00 | 100.00 | 0.00 | 0.00 |

and a positive value of over 7 percentage points in the latter. The built-up area during this period rose substantially by over 4 square kilometres.

The built-up area constituted just over 10 per cent of the city area in 1990 and increased to 16.8 per cent by 1997. However, with increasing slope, generally the proportion of built-up area does come down and forest area does record an increase (Table 4.4, see also Figures 11). In each slope-class, the largest proportion of area is under forest of various types. Understandably, the proportion of built-up area is low in areas with a slope exceeding 50 degrees and it occupies small patches of habitation. Land with slope between 30 to 50 degrees also records substantial proportion of built-up area. Areas not utilised for construction and located in the slope category of less than 30 degrees (Figure 12) would require further analysis regarding their ownership and other attributes.

If we look at each land use category, the highest proportion of land in each is found in areas with over 50 degrees of slope, since such area accounts for over 76 per cent of the city. Over 50 per cent of the built-up area is found in land with more than 50 degree slope.

Given this picture, if any land is to be converted to built-up area one needs to seek parcels of land in areas classified under 30 degree of slope and lying barren which amounts to just about 1.24 square kilometres in 1997. Analysis at a more micro scale may reveal small parcels of land with lower degrees of slope than indicated in this study.

Another way of analysing the pattern of conversion of land is to create a matrix which tells us how much of each category of land in the base year was converted and to what use it was put over a time span (1990 to 1997 in this study). Such a matrix is given in Table 4.5. A study of this table reveals that: (a) of about 34 square kilometres of area under dense forest in 1990, only 15.5 remained as dense forest in 1997. Part of it was reduced to the status of open forest (9.08 square kilometres) and about 4.52 square kilometres were converted as built-up area, (b) of about 11 square kilometres that were originally classified as open forest a little over

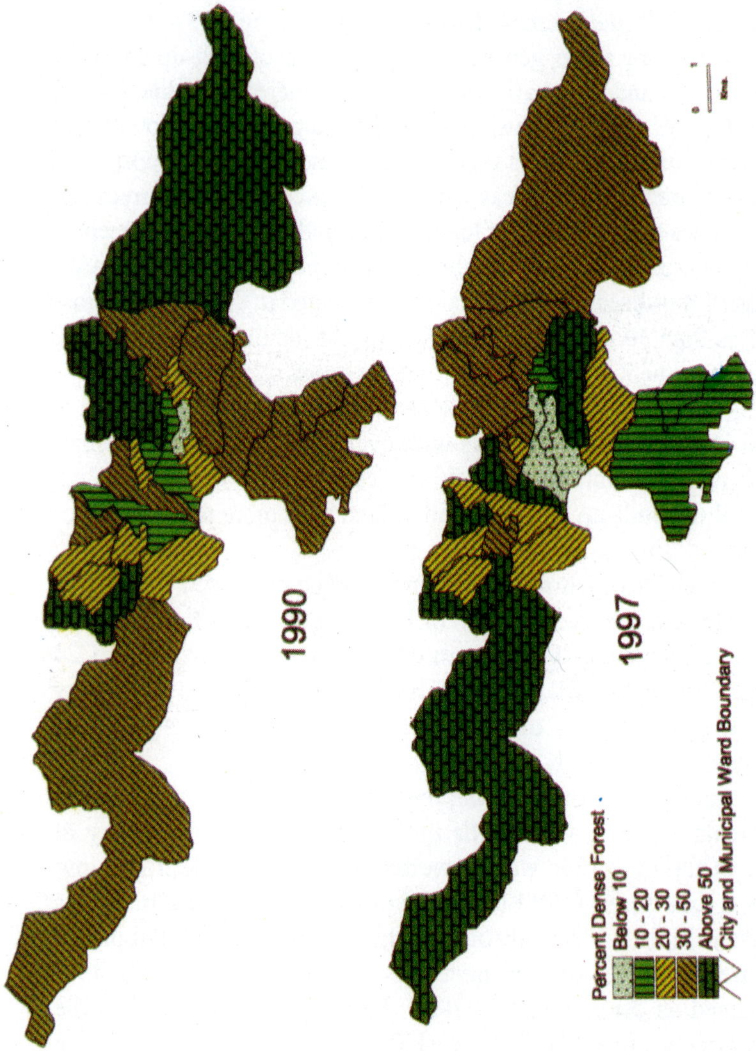


**Fig. 9 :** Mussorie—Per cent Area Under Dense Forest

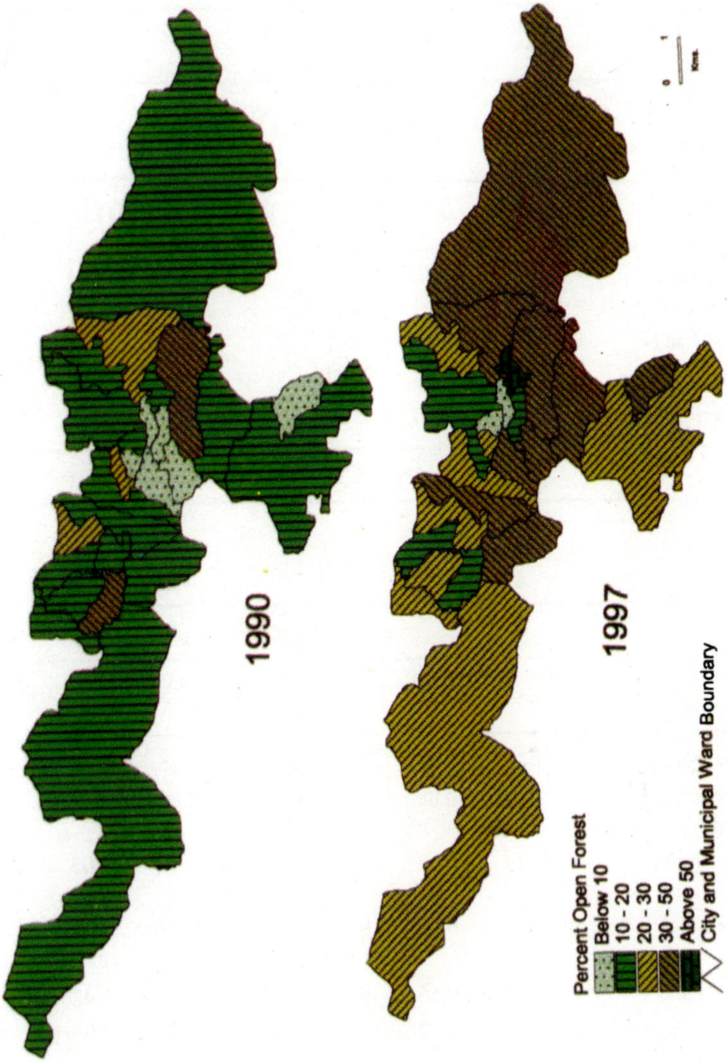


**Fig. 10 :** Mussoorie—Per cent Area Under Open Forest

**Table 4.3:** Use of Land in Mussoorie by Area under Various Categories of Slopes, .1990.

*(in square kilometres)*

| Land Use Categories | Degree of Slope | | | | | |
|---|---|---|---|---|---|---|
| | Below 10 | 10-20 | 20 -30 | 30-50 | Above 50 | Total |
| Dense Forest | 2.41 | 1.58 | 1402 | 3.90 | 20.03 | 29.32 |
| Medium/Open Forest | 1.67 | 0.35 | 0.54 | 1.73 | 6.79 | 11.08 |
| Scrub/Shrub Land | 0.61 | 0.72 | 0.82 | 1.23 | 8.24 | 11.62 |
| Agricultural Land | 0.14 | 0.14 | 0.23 | 0.50 | 1.50 | 2.51 |
| Barren Land | 0.37 | 0.34 | 0.47 | 0.97 | 2.40 | 4.55 |
| Built-up Area | 0.34 | 0.53 | 0.77 | 2.10 | 2.99 | 6.74 |
| Unclassified | 0.00 | 0.00 | 0.00 | 0.00 | 0.00 | 0.00 |
| Total | 5.54 | 3.65 | 4.23 | 10.43 | 41.95 | 65.81 |

In Sq. Km. - 1997

| Land Use Categories | Degree of Slope | | | | | |
|---|---|---|---|---|---|---|
| | Below 10 | 10-20 | 20 -30 | 30-50 | Above 50 | Total |
| Dense Forest | 1.40 | 0.91 | 1.35 | 3.51 | 16.15 | 23.32 |
| Medium/Open Forest | 1.39 | 0.88 | 1.01 | 3.02 | 11.98 | 18.28 |
| Scrub/Shrub Land | 0.76 | 0.48 | 0.55 | 1.08 | 3.18 | 6.05 |
| Agricultural Land | 0.18 | 0.11 | 0.14 | 0.42 | 2.72 | 3.57 |
| Barren Land | 0.58 | 0.56 | 0.10 | 0.22 | 0.96 | 2.42 |
| Built-up Area | 1.23 | 0.72 | 1.07 | 2.11 | 5.93 | 11.06 |
| Unclassified | 0.00 | 0.00 | 0.00 | 0.07 | 1.03 | 1.10 |
| Total | 5.54 | 3.65 | 4.23 | 10.43 | 41.96 | 65.81 |

**Table 4.3** *(contd.)*

In Sq. km. Total change

| Land Use Categories | Degree of Slope | | | | | |
|---|---|---|---|---|---|---|
| | Below 10 | 10-20 | 20 -30 | 30-50 | Above 50 | Total |
| Dense Forest | -1.02 | -0.67 | -0.05 | -0.39 | -3.88 | -6.01 |
| Medium/Open Forest | -0.28 | 0.53 | 0.47 | 1.29 | 5.19 | 7.20 |
| Scrub/Shrub Land | 0.15 | -0.24 | -0.27 | -0.15 | -5.06 | -5.57 |
| Agricultural Land | 0.04 | -0.03 | -0.09 | -0.08 | 1.22 | 1.06 |
| Barren Land | 0.21 | 0.22 | -0.37 | -0.75 | -1.44 | -2.13 |
| Built-up Area | 0.89 | 0.19 | 0.30 | -0.01 | 2.94 | 4.33 |
| Unclassified | 0.00 | 0.00 | 0.00 | 0.07 | 1.03 | 1.10 |
| Total | 0.00 | 0.00 | 0.00 | 0.00 | 0.00 | 0.00 |

5 square kilometres appear to have become densely forested and (c) a large part of the built-up area seems to have been obtained from densely forested area. These figures are approximate, largely because of the resolution of imagery used for the two periods of time and must not be taken as exact values.

**Table 4.4:** Use of Land in Mussoorie—Per cent Area under Various Land use Categories by Degree of Slope

(Percentage Area to Total of the Slope - 1990)

| Land Use Categories | Degree of Slope | | | | | |
|---|---|---|---|---|---|---|
| | Below 10 | 10-20 | 20-30 | 30-50 | Above 50 | Total |
| Dense Forest | 43.53 | 43.14 | 33.10 | 37.37 | 47.74 | 44.55 |
| Medium/Open Forest | 30.16 | 9.64 | 12.77 | 16.61 | 16.17 | 16.84 |
| Scrub/Shrub Land | 11.01 | 19.58 | 19.39 | 11.84 | 19.65 | 17.66 |
| Agricultural Land | 2.48 | 3.75 | 5.49 | 4.75 | 3.58 | 3.81 |
| Barren Land | 6.63 | 9.30 | 11.14 | 9.26 | 5.73 | 6.91 |
| Built-up Area | 6.19 | 14.62 | 18.11 | 20.17 | 7.13 | 10.24 |
| Unclassified | 0.00 | 0.00 | 0.00 | 0.00 | 0.00 | 0.00 |
| Total | 100.00 | 100.00 | 100.00 | 100.00 | 100.00 | 100.00 |

(Percentage Area to Total of the Slope - 1997)

| Land Use Categories | Degree of Slope | | | | | |
|---|---|---|---|---|---|---|
| | Below 10 | 10-20 | 20 -30 | 30-50 | Above 50 | Total |
| Dense Forest | 25.20 | 24.87 | 31.99 | 33.62 | 38.49 | 35.43 |
| Medium/Open Forest | 25.17 | 24.06 | 23.92 | 28.95 | 28.55 | 27.78 |
| Scrub/Shrub Land | 13.69 | 13.08 | 13.10 | 10.39 | 7.58 | 9.20 |
| Agricultural Land | 3.24 | 2.95 | 3.40 | 4.02 | 6.49 | 5.43 |
| Barren Land | 10.48 | 15.22 | 2.35 | 2.08 | 2.29 | 3.67 |
| Built-up Area | 22.22 | 19.82 | 25.25 | 20.24 | 14.14 | 16.81 |
| Unclassified | 0.00 | 0.00 | 0.00 | 0.71 | 2.46 | 1.67 |
| Total | 100.00 | 100.00 | 100.00 | 100.00 | 100.00 | 100.00 |

**Table 4.4:** *(Contd.)*

(Percentage Area to Total Landuse Category—1990)

| Land Use Categories | Degree of Slope | | | | | |
|---|---|---|---|---|---|---|
| | Below 10 | 10-20 | 20 -30 | 30-50 | Above 50 | Total |
| Dense Forest | 8.22 | 5.38 | 4.78 | 13.29 | 68.33 | 100.00 |
| Medium/Open Forest | 15.08 | 3.17 | 4.88 | 15.63 | 61.25 | 100.00 |
| Scrub/Shrub Land | 5.25 | 6.16 | 7.06 | 10.62 | 70.91 | 100.00 |
| Agricultural Land | 5.48 | 5.48 | 9.29 | 19.76 | 60.00 | 100.00 |
| Barren Land | 8.08 | 7.48 | 10.37 | 21.23 | 52.83 | 100.00 |
| Built-up Area | 5.09 | 7.93 | 11.38 | 31.21 | 44.40 | 100.00 |
| Unclassified | 0.00 | 0.00 | 0.00 | 0.00 | 0.00 | 0.00 |
| Total | 8.42 | 45755 | 6.43 | 15.84 | 63.75 | 100.00 |

(Percentage Area to Total Landuse Category—1997)

| Land Use Categories | Degree of Slope | | | | | |
|---|---|---|---|---|---|---|
| | Below 10 | 10-20 | 20 -30 | 30-50 | Above 50 | Total |
| Dense Forest | 5.99 | 3.90 | 5.81 | 15.04 | 69.27 | 100.00 |
| Medium/Open Forest | 7.63 | 4.81 | 5.54 | 16.51 | 65.52 | 100.00 |
| Scrub/Shrub Land | 12.53 | 7.89 | 9.16 | 17.89 | 52.53 | 100.00 |
| Agricultural Land | 5.03 | 3.02 | 4.02 | 11.73 | 76.21 | 100.00 |
| Barren Land | 24.05 | 23.02 | 4.11 | 8.99 | 39.82 | 100.00 |
| Built-up Area | 11.13 | 6.55 | 9.65 | 19.07 | 53.60 | 100.00 |
| Unclassified | 0.00 | 0.00 | 0.00 | 6.71 | 93.89 | 100.00 |
| Total | 8.42 | 5.55 | 6.43 | 15.85 | 63.76 | 100.00 |

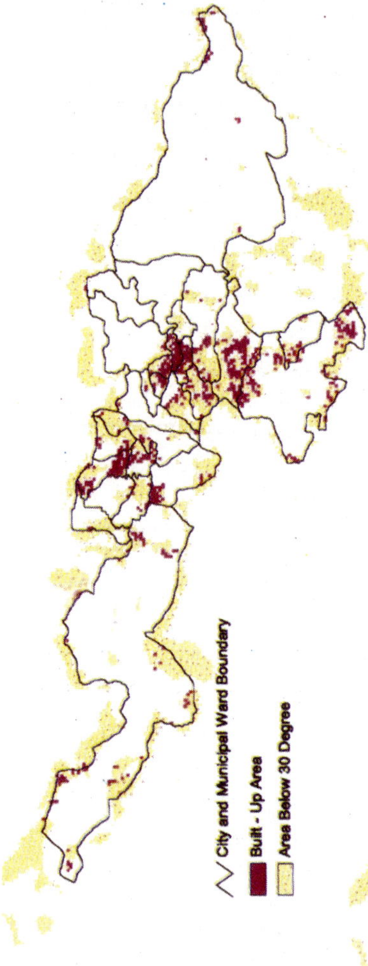


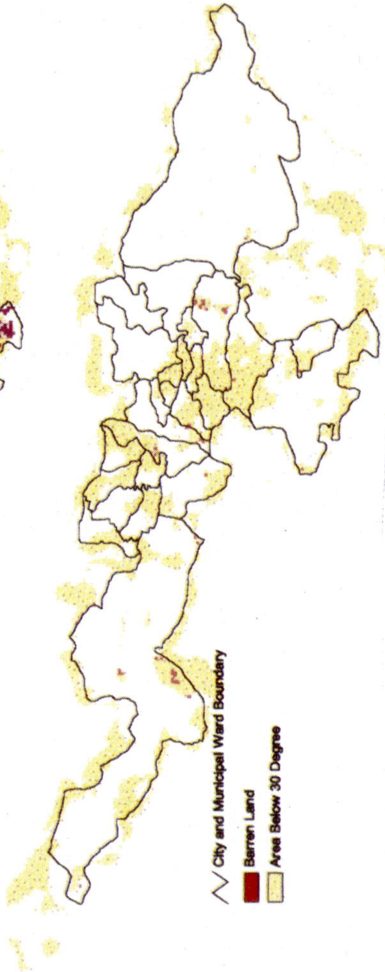


**Fig. 12A:** Mussoorie—The Use of Land: Area Below 30 Degree Slope Built-up Area

**Fig. 12B:** Mussoorie—The Use of Land: Barren Land

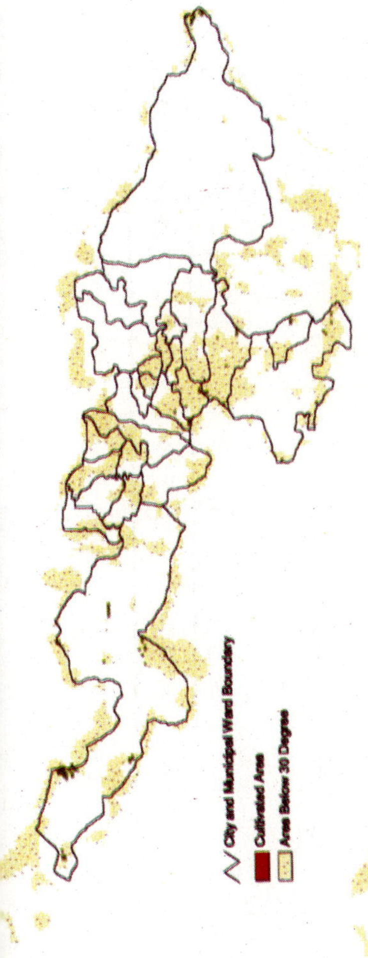


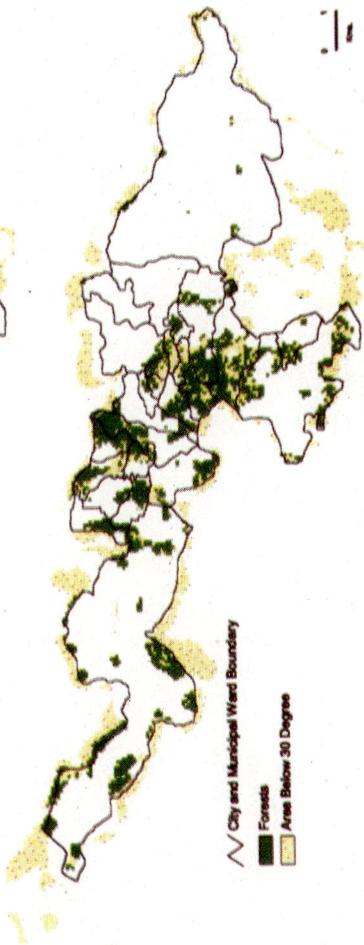


**Fig. 12C:** Mussoorie—The Use of Land: Cultivated Area

**Fig. 12D:** Mussoorie—The Use of Land: Land Forests

**Table 4.5:** Mussoorie Land Conversion Matrix - 1990—97

*(in square kilometres)*

| 1997 \ 1990 | Dense Forest | Medium/ Open Forest | Scrub/Shrub Land | Cultivated Land | Barren Land | Built-up Area | Uncla-ssified | Total |
|---|---|---|---|---|---|---|---|---|
| Dense Forest | 15.55 | 9.08 | 2.59 | 1.18 | 0.25 | 4.52 | 0.72 | 33.89 |
| Medium/ Open Forest | 5.41 | 3.72 | 0.48 | 0.23 | 0.28 | 1.02 | 0.22 | 11.36 |
| Scrub/Shrub Land | 2.00 | 4.37 | 2.22 | 0.70 | 0.42 | 2.11 | 0.09 | 11.91 |
| Agricultural Land | 0.24 | 0.69 | 0.55 | 0.38 | 0.02 | 0.57 | 0.02 | 2.47 |
| Barren Land | 0.11 | 0.42 | 0.21 | 0.21 | 1.24 | 0.28 | 0.00 | 2.46 |
| Built-up Area | 0.00 | 0.00 | 0.00 | 0.87 | 0.19 | 2.54 | 0.05 | 3.65 |
| Total | 23.31 | 18.28 | 6.05 | 3.57 | 2.40 | 11.04 | 1.10 | 65.75 |

# 5

# Concluding Observations

The preceding analysis may be treated at one level as a micro study involving residents, their representatives in the Municipal Board and other interest groups. At another level, it is more generalised, since study of the physical aspects involved using map and imagery information and maps on a scale and resolution that is, perhaps, inadequate for the micro urban context. However, on the basis of the analysis we may come to some tentative conclusions.

- Firstly, the city residents are aware of the developments and deterioration of living conditions over time and have been able to rank the city's problems as they perceive them. The participatory approach resulted in identification of major concerns:

  - Inadequate land base of the city for residential requirements of the permanent population in the background of rapid expansion of built-up area in the nineties, largely for commercial purposes and speculative land holding.
  - The dilemma of the need to protect and strengthen the tourist industry while minimising negative impacts of such growth manifested in pressure on services and city infrastructure including water, sanitation, land for housing, power, transportation and congestion.
  - The overriding role of development agencies that tend to de-empower elected representatives at the grassroots, dated regulations and bye-laws that govern city development, the insensitive manner of implementation of regulations,

and lack of a development plan for the city, are important factors that emerge clearly from the preceding analysis. One of the immediate requirements is to involve the elected members of the City Board in urban development decisions.

- The participatory processes adopted in this study brings out clearly that it is difficult to convince a substantial section of the resident population, that the city has grown beyond the threshold of sustainability when on many considerations—most of all the national/public interest considerations and commercial considerations–they see the city expanding, yet they find it difficult to get permission to even repair their ageing shelters, or to put up an additional room to house the expanding family. The position of technocrats and planners is even more untenable, when we find that permission for construction is often denied since most part of the city is classified as forest, without defining what constitutes a forest. One of the responsibilities of the Development Authority is to increase the level of awareness of the residents with reference to current status regarding permissions and approvals required for repair and maintenance of old structures.

- Secondly, the study has shown poor and grossly inadequate database on which development activities are planned. It has been argued earlier in this study that there is a need to include temporal considerations in estimating even such basic parameters as population of the city. Often resource allocations are based on per-capita basis and in cities like Mussoorie it becomes necessary to estimate tourist population, which in several months of the year far exceeds the resident population.

- Thirdly, based on standard norms and requirements it has been pointed out that:

- The requirement of water far exceeds the availability even at present for nine months of the year, the three months of adequate water supply being in those winter months when a third of the resident population is out of Mussoorie. The projections also show that by the year 2016 current water supply will be inadequate even during the winter months.
- Estimates of waste generation (based on trials conducted separately for resident and tourist population) indicate the vast difference in the waste generated between the peak season and the lean season—with the former recording three times that of the latter. The increase during the next two decades is estimated to be one and half times the present. The fact, that tests conducted show that the water quality in and around Mussoorie is within the pollution limits prescribed has been noted. However, with a single dumping ground whose location poses a major threat of polluting one of the important sources of drinking water of the city, waste management remains a problem to be attended to on a priority basis.
- The demand for tourist accommodation is again to be seen in the context of temporal dimension. The supply exceeds the demand by two times during nine months in a year and is likely to be so even in the next quarter of a century, while substantially short during the peak season. The answer obviously, is not in creating/augmenting capacities in hotels; the problem begs a more innovative solution.
- Medical facilities expressed in terms of availability of hospital beds indicate a shortfall even at present during the peak season and it is expected to be so throughout the year in the next two decades at the current level of supply. As noted earlier, availability of land will be the limiting factor in expansion of land dependent infrastructure.
- The capacity for distribution of electric power could be adequate with a small investment in erecting additional sub-stations and transformers, but the captive power supply is extremely limited and the bulk of the demand is met

through feeder lines. The residents have witnessed 4 to 8 hours of load shedding during winter, when demand for energy for heating is acute. This doubtless results in the use of other sources of energy including firewood from city forests.

- Fourthly, the analysis has shown that practically no land that is barren and located on gentle slopes is available for future development. About 4,300 hectares of land was used for construction in a period of about eight years between 1990 and 1997. The total built area up to 1990 was 6,740 hectares. Such rapid expansion owing to commercial pressure has tended to keep the resident population outside the land market.

- Fifthly, a substantial proportion of land with high degree of slope has been put to urban uses particularly during the previous eight years. That the construction cannot expand is clear, but a solution is needed for the housing demands of the resident population. Although most sections of the resident population believe that the Development Agencies must be kept out of the city limits, it can achieve very little in so far as urban development within the city limits is concerned in the background of the burden of rapid expansion that left the city bereft of any further scope of expansion.

# ANNEXURES

*ANNEXURE - 1*

## Water Supply and Sewerage System in Mussoorie

### Sources of Water

Owing to its mountainous terrain, sources of water for Mussoorie are limited to mountain streams, brooklets and rainwater, and do not include ground water. Water available from the first two sources has to be lifted over heights most of the time. Water supply, therefore, is dependent on uninterrupted and steady supply of electricity. Water available from harvesting rainwater is not completely clean and, therefore, has limited use.

Ironically, water availability is least when it is needed most. This is true for every place but in the case of Mussoorie it has unique characteristics. During summer months demand for water goes up not only because of rising temperatures but also because of increased population load of tourists. At the same time, frequency of rains and discharge of streams fall. Electricity supply becomes more erratic and power cuts more frequent.

At present water for Mussoorie is being drawn from 17 sources, 14 of which are springs and three are brooklets. Of the 17 sources, water has to be pumped up from 12 and from the remaining sources, gravity is used to carry water to the residents. Water discharge of 12 sources, from which water is carried up by pumping is 6,021 litres per minute (lmp), or 8.67 million litres per day (mlpd). Water discharge of the remaining five sources is 360 litres per minute (lmp), or 0.52 million litres per day (mlpd). Total water discharge is 6,381 lmp or 9.19 mlpd. Details of sources of water and demand as projected by *Jalsansthan* are given in Tables 1, 2 and 3. Rain,

**Table 1:** Garhwal Jal Sansthan :
Mussoorie Water Supply Scheme/Total Water availability

| Sl. No. | Name of Water Source | Nature of Source | | Discharge of source (lpm*) | | System of W/S | |
|---|---|---|---|---|---|---|---|
| | | Spring | Brook | Spring | Brook | Pumping | Gravity |
| 1. | Murrey Pump | | | | | | |
| (i) | Khanlty | spring | — | 450 | — | pumping | — |
| (ii) | Under Cliff | spring | — | 108 | — | do | — |
| (iii) | Bansi or Lyndale | spring | — | 280 | — | do | — |
| 2. | Khandighat (old) | — | Brooklet | — | 336 | do | — |
| (i) | Khandighat 30 H.P. Lower | spring No. 8 | — | 225 | — | do | — |
| (ii) | Khandighat, New 25 H.P. | | | | | | |
| 3. | John Mackinon | | | | | | |
| (i) | New-Bee | spring | — | 50 | — | do | — |
| (ii) | John Mackhinon | spring | — | 155 | — | do | — |
| (iii) | Chamber khud | spring | — | 32 | — | do | — |
| 4. | Bhilaru Pump | | | | | | |
| (i) | Bhilaru upper spring, 100 BHP | spring | — | 1700 | — | do | — |
| 5. | Jinsy Pump | | | | | | |
| (i) | Jinsy stage-2 | spring | — | 400 | — | do | — |
| (ii) | Jinsy stage-1 | — | Brooklet | — | 1680 | do | — |
| 6. | Kolty Khala | | | | | | |
| (i) | Kolty Khala | — | Brooklet | — | 605 | do | — |
| 7. | Gravity Source | | | | | | |
| (i) | Company Khud | spring | — | 40 | — | — | gravity |
| (ii) | Brook Land | spring | — | 90 | — | — | do |
| (iii) | Nalapani | spring | — | 30 | — | — | do |
| (iv) | Pargakhala | spring | — | 110 | — | — | do |
| (v) | Duglusdail | spring | — | 90 | — | — | do |
| | **Total** | **14 Nos.** | **3 Nos.** | **3760** | **2621** | **12 Nos.** | **5 Nos.** |

*Note:* Total Water availability = 3760+2621 = 6381 lpm.
* lpm (litres per minute)

which is frequent in hills, is another source of water. Most of it either seeps into the ground or runs off down the slope. Some amount is, however, harvested using traditional practices. Houses in Mussoorie have sloping roofs. Rain water falling on the roof slides down into an open pipe fixed at the base of the roof. This pipe is tilted to one side. Water falling into it thus flows to that side and into a tank placed on the ground directly below this end of the pipe. The tank is placed either on the surface or is underground. Water can be stored in these tanks for long periods, to be used as and when needed.

*Jal Sansthan,* Mussoorie, has water storage capacity of 16 m.l. (million litres) in balancing reservoirs at Gunhill. At times when there is a surplus, excess water is pumped into these storage tanks to be used during days of peak demand.

## Supply of Water

*Jal Sansthan,* Mussoorie, pumps up 6.5 to 7 mlpd (million litres per day) of water from water sources like springs and brooklets in and around Mussoorie. In addition to this, about 0.51 mlpd water is available, which is collected and supplied using gravity and for which no pumping is required. The latter is made available to supply to areas that are at a lower altitude than the water sources. They are Barlowganj, Jharipani, Bhatta, Kolhukhet and Bala Hissar. These areas receive water 24 hours of the day.

The remaining parts of Mussoorie are supplied water by pumping, since they lie at a higher altitude than the water sources. Here water supply is rationed daily: two hours each in the morning and evening. During summers, supply is sometimes increased to 2.5 hours in the morning, depending on availability. Water supply using pumps depends directly on electricity supply and the supply voltage. Power cuts of three to four hours result in a drop in the supply by 1-2 mlpd. During summer months, when demand is highest, electricity supply is most erratic. *Jal Sansthan* has taken some measures to counter the effect of low voltage supply, by increasing the pumping stages in one pipeline from 9 to 10 pumps.

**Table 2:** Demand for Water of Permanent Population of Mussoorie[6]

| | Peak Load | Average Load |
|---|---|---|
| Permanent Population | 40,000 | 34,000 |
| Water demand (mlpd @ 175 lpcd) | 7 | 5.95 |
| Water demand (mlpd @ 120 lpcd) | 4.8 | 4.08 |

**Table 3:** Demand for Water of Total Population of Mussoorie[7]

| | Peak Load | Average Load |
|---|---|---|
| Total Population | 100,000 | 80,000 |
| Water demand (mlpd @ 175 lpcd) | 17.50 | 14.00 |
| Water demand (mlpd @ 120 lpcd) | 12.00 | 9.60 |

Water stored in balancing reservoirs at Gunhill is used to meet shortfalls in water supply during the summer months. Besides the storage capacity of *Jal Sansthan*, individual hotels, resorts, cottages, houses, etc. have their private storage capacity to tide over times of low supply and high demand. No data regarding this capacity is available. It is estimated, however, that the only 5-star Hotel in Mussoorie, Jaypee Resorts has a storage capacity of nearly 0.5 m.l., and the Lal Bahadur Shastri National Academy of Administration nearly 0.2 m.l.

[6] The population figure of 34,000 is the figure considered by Jal Sansthan, Mussoorie, in estimation of the water demand in Mussoorie. The permanent population figure of 40,000 for MSR, made available by City Board, Mussoorie, however appears to be more realistic. Water supply on the basis of 175 litres per capita per day (lpcd) consumption is considered adequate by the *Jal Sansthan* to enable a person to wash clothes and bathe twice in the day. Water supply based on 120 lpcd consumption level is considered by the *Jal Sansthan* as the minimum consumption level. During peak tourist season, Jal Sansthan tries to maintain water supply at least at this level.

[7] During the peak tourist season, population in Mussoorie touches 1 lakh, comprising tourists and residents of Mussoorie. The peak population load of 1 lakh is confined to weekends, during weekdays the load is less. The average load thus works out to be 80,000. Maximum and minimum water consuption levels of 175 Ipcd and 120 Ipcd are considered.

## Proposed and Ongoing Works

### *Works Completed*

Two storage tanks with a capacity of 4 ml each, at Radha Bhawan and Craig Top are expected to be handed over to *Jal Sansthan* by March 1998. This will increase water storage capacity by 8 ml.

### *Works in Progress*

Work on a new pumping scheme to lift 1 mlpd of water from Dhobhighat to Library, started in 1997. It is expected to be completed by 1999-2000.

### *Works Proposed*

Aglad Water Pumping Scheme to pump 12 mlpd water from Aglad River. The scheme which was estimated to cost Rs. 7 crores in 1984 was shelved, partly because of opposition from farmers drawing water from the river for farming purposes. In 1998 it is estimated to cost upto Rs. 20 crores.

## The Sewerage System

The sewerage system in Mussoorie was designed by the British and installed in 1934. The system was planned for 50 years, i.e. till the year 1984. The total length of sewer lines is approximately 21 kilometres and there are 1,100 manhole chambers. There are eight sewerage zones, and a total of 4,225 sewer seats, as on September 1997. Sewer collected in each zone is stabilised in a septic tank and soak pit, located at the terminus of sewer lines in that zone. Sewer lines are made of Cast Iron Socket and Spigot (CISNS) pipes; and pipe diameters are 100 mm, 150 mm, 200 mm.

Mussoorie's hilly terrain offers some advantages to the system. Flow of sewerage through pipes is much easier due to natural gradients. Hence pipes with smaller diameters are also effective. There is no need for pumping the waste and blockages are fewer.

At the same time, Mussoorie's sewerage system suffers some disadvantages. The system is old and has exceeded its life span. The original layout map of the sewer lines does not exist any more. Some sewer lines have been disconnected from their natural outfalls and connected to other outfalls. There is no effective drainage system in Mussoorie. Thus, rain water also flows into the sewer lines. Kitchen drainage is often illegally connected to sewer lines. Kitchen waste sometimes clogs the lines. Sewer outlets of illegal constructions are joined to the sewer lines. Excess construction material like rubble, sand, etc. is sometimes dumped into sewer lines.

**Quantity of Sewerage Generated**

Almost 75 per cent of the total water supplied to the town reaches its sewer lines. The remaining water is lost by seepage into soil, surface runoff and evaporation. Since there is no estimate of the quantity of rain water and snow, only load on the sewer lines from water supplied by the Municipality can be estimated.

| | Average | Peak |
|---|---|---|
| Water supply (mlpd) | 7.00 | 9.00 |
| Sewerage generated (mlpd) (@ 75%) | 5.25 | 6.75 |

**Sewerage Layout**

The sewerage system in Mussoorie caters only to some parts of the town. The eight zones into which it is divided do not cover the entire town. Also, there are residential establishments within these zones that are not connected to the sewer lines. Such establishments have alternative methods of sewerage disposal like individual septic tanks and soak pits and use of the conservancy system of disposal.

For the remaining parts of the town, sewerage disposal is through a network of sewer lines terminating in septic tanks and soak pits. Sewer lines, carry sewage from houses, hotels, resorts, schools, institutions, etc. and are connected to the sewer system.

At the terminus of each zone is a stabilisation pond comprising of a septic tank and a soak pit. Sewage drains first into the septic tank. Sludge, or the solid content, settles in this tank. Overflow from the septic tank flows into the soak pit. Part of the water gets soaked here and the remaining part drains out. Some of the stabilisation ponds are made of three chambers in place of two, sewerage stabilisation then is better and water draining out from them is clearer. The water draining out from the stabilisation pond is not used further. The volume of sludge that accumulates in the septic tank is very little because the solid matter is well mashed before it reaches the septic tank due to the natural gradient available in Mussoorie. Hence, cleaning of septic tanks is required when it is almost full, i.e. once in 10 years.

*Jal Sansthan*, Mussoorie prepared and presented a feasibility report of the Mussoorie Sewerage Scheme in May, 1989. The proposed works were in appreciation of the increasing load on the sewerage system in Mussoorie. There was also an understanding that the system was functioning well beyond its life span.

Areas where the sewerage system does not exist are:

1. Lal Tibba Cantonment area at Mussoorie, Tehri Bus Stand, Wood Stock, Jabarkhet, Bataghat.
2. Landour behind Clock Tower via Rajpur Road, Wynberg Allen School, below Nanpara, Bala Hissar upto Chandra Bhawan, upper area at Barlowganj.
3. Below Jain Dharmshala, Landour up to Picture Palace on Landour Road.
4. Masonic Lodge Bus Stand upto King Craig main road, upper and lower areas also, upto Kolukhet.
5. Camel's Back Road, below Mount Rose up to Camel's Back Road cemetry.
6. Library Chowk to Nandatels, Library Kingcraig Road.
7. Hotel Hakmans up to Vasu Cinema on the Mall Road.
8. Spring Road and Total Circular Road, Radha Bhawan.
9. Dick Road, Harnamsingh Road upto Company Garden area below the road.

10. From Srinagar Estate onwards upto Kempty Fall Road.
11. Library to Charleville Road, just below Kempty Bus Stand.
12. Dhumanganj, Hathipaon, Sangrilla upto Cloud's End Resort.

The study was commenced in 1983 when the population of Mussoorie was estimated to be more than 30,000. The new scheme was designed for the next 50 years, i.e. until 2023. The population projection for 2023, which has been used as the design population, is 1,20,000 for permanent population and 2,00,000 for the total urban population load (including tourists load at any point of time). The scheme envisages creating nine additional sewerage zones and works include replacement and laying of new lines and manholes.

## Concluding Remarks

Mussoorie has the capacity to supply 7 to 7.5 mlpd of water. Of this, 0.51 mlpd that is supplied all 24 hours to low lying areas by the use of gravity, is sufficient for those areas. The remaining water supplied by pumping depends greatly on electricity supply and supply voltage. Measures taken by the *Jal Sansthan* like increasing the number of pumping stages can only be a partial solution. During the winter season when tourist pressure is low, permanent residents including boarding school children move down to the plains and water demand per capita comes down, water supply is adequate. During winters bathing is restricted to a maximum of one time daily and washing is minimal, per capita water demand thus is in the range of 120 lpcd. For the permanent population estimate of 40,000 (as estimated by the *Jal Sansthan*), total water demand is 4.8 mlpd. This is almost 2 mlpd less than the water supply capacity of 7 mlpd. Excess water available during the winter is collected in the balancing reservoirs. Increased demand over weekends, which is experienced mostly by hotels and resorts, is met by their private storage tanks in which excess water supply during weekdays is collected. The two storage tanks at Craig Top which are ready for handing over to *Jal Sansthan* will increase the latter's reservoir capacity by 8 ml. This would be a valuable addition for summers.

The third tank at Craig Top, which is under construction, will further enhance reservoir capacity by 4 m.l. The new water pump scheme from Dhobighat to Library will increase water supply by 1 mlpd. In the light of the expected increase in population pressure in Mussoorie, implementation of the Algad Water Pumping Scheme is essential to cater to the increased load.

The Mussoorie sewerage system was designed for the period upto 1984. The system has already outlived its life by over 13 years. There are problems of leaking and damaged lines, stolen C.I. pipes and damaged septic tanks. Pipelines are laid in the open and are vulnerable to damage by falling mountain rocks. CISNS pipes are very expensive, i.e. Rs. 1,000 per meter. Hence replacement of existing lines and laying new lines would involve a large expense. The lines have to bear additional load for which no provision was made in the design. This includes illegal connection of kitchen drainage and sewer lines of illegal constructions. In the absence of a proper drainage system in Mussoorie, rain water also flows into the sewer lines. This often exceeds the capacity of the latter thereby leading to overflow of manholes during the monsoons. Though the registered number of seats is 4,225, the *Jal Sansthan* estimates the actual number to touch almost 5,000.

Disposal of construction material down the sewer lines leads to frequent blockages. Hotels often construct private septic tanks/ soak pits without prior permission and capacity calculation. Hence standard norms of construction are sometimes violated leading to problems in the future. Absence of layout plans of the sewerage system often leads to problems in repairs and maintenance. The existing system caters to only 50 per cent of the town area, leaving the remaining areas to make provisions for sewerage disposal by individual and private methods.

**Table 4:** Zonal Division of Sewage System in Mussoorie

| Sl. No. | Name of Zone | Name of Outfall | No. of Sewer Seats | Line Diameter (Inches) | Areas Covered | Problems | Works Undertaken or in Progress |
|---|---|---|---|---|---|---|---|
| 1. | Landour | Butcher Khana | 407 | 4,6 | Landour Bazar, South Road, North Road Butcher Khana from Nanpara to North Road. | Line damage, manhole repair | Petty repairs |
| 2. | Barlowganj | Barlowganj | 24 | 4,6 | Below Jaypee Hotel gate, Barlowganj Bazar | Line damage, other repairs | None |
| 3. | Kulri | Sarai | 1,319 | 4,6 | Hotel Shipra via Picture Palace, Mayfield, Ram Bhawan upto Sarai septic tank, Below Rockwood, Kulri Mall Road., Jafarhall, Victory House upto Hira Bhawan, Below Hotel Deep Camel's Back Road, upto Sarai | Septic tank, wall damaged, CI pipes Stolen | None |
| 4. | Bhilaru | Bhilaru | 1,019 | 6,8 | Chaman Estate, Motilal Nehru Road, behind Library Bazar, Hotel Oasis, Mefel Hedge, Indra Bhawan, PWD Inspection House, Ekant, via Srinagar Estate | None | 4 Suspension bridges repaired |
| 5. | Arcadia | Arcadia | 220 | 4,5 | Hotel Glenvilla, via Tar Gali, Peak View Hotel, Pushp Vatika, Dila Ram | Septic tank damaged CI pipes stolen | None |

**Table 4:** *(Cont.)*.....

| Sl. No. | Name of Zone Seats | Name of Outfall (Inches) | No. of Sewer | Line Diameter | Areas Covered or in Progress | Problems | Works Undertaken |
|---|---|---|---|---|---|---|---|
| 6. | Shergarhi | Shergarhi | 716 | 4,5,6 | Kulri Mall Rd. via Hotel Samrat, Stawari Bank, Sussex House, Cliff Hall Estate, Hussain Ganj, Murad Ganj, Rialto Gate, Naj Gali, Hampton Court | Manhole damaged, minor repairs | Mahole repair in progress |
| 7. | Happy Valley | Happy Valley | 508 | 6 | LBSNAA main gate to Happy Valley, Kempty Road upto Kali Dang | Septic tank & soak pit damaged, CI pipes | None |
| 8. | Company Bagh | Company Bagh | 12 | 6 | Hathi Paon Road upto Company Bagh Septic tank | Septic Tank damaged, minor repairs | None |

**Table 5:** Mussoorie Sewerage Scheme —Proposed Sewer Line in Different Zones

*(Line length in meters)*

| Sl. No. | Zone | Existing Sewer Lines | | | | Proposed Sewer Lines | | | | | New Sweer Lines to be Laid | | | | | Manhole |
|---|---|---|---|---|---|---|---|---|---|---|---|---|---|---|---|---|
| | | 100 mm dia | 125 mm dia | 150 mm dia | 200 mm dia | 150 mm dia | 200 mm dia | 250 mm dia | 300 mm dia | 350 mm dia | 150 mm dia | 200 mm dia | 250 mm dia | 300 mm dia | 350 mm dia | (Slab type, size 1*1.20 1.50 m deep) |
| 1. | Company Garden | 1500 | - | - | - | 1500 | 500 | - | - | - | 1500 | 500 | - | - | - | 135 |
| 2. | Happy Valley | 918 | - | 2900 | - | 1000 | 1000 | 1000 | 1500 | - | - | 1000 | 1000 | 1500 | - | 235 |
| 3. | Bhilaru | 658 | 996 | 376 | 1400 | 500 | 1000 | 500 | 800 | - | 300 | 300 | 500 | 800 | - | 125 |
| 4. | Scandal Point | - | - | - | - | 1000 | 1000 | - | - | - | 1000 | 1000 | - | - | - | 135 |
| 5. | Cemetry | - | - | - | - | 1000 | 1100 | - | - | - | 1000 | 1100 | - | - | - | 1400 |
| 6. | Arcadia | 1280 | - | - | - | 1000 | 800 | - | - | - | 1000 | 800 | - | - | - | 120 |
| 7. | Sarai | 2147 | - | 724 | - | 2000 | 100 | 500 | 500 | 500 | 1500 | 1000 | 500 | 500 | 500 | 265 |
| 8. | Butcher Khana | 504 | - | 1000 | - | 1200 | 800 | 400 | 700 | - | 700 | 800 | 400 | 700 | - | 170 |
| 9. | Cremation Ground | - | - | 950 | - | 600 | 400 | 500 | 900 | - | - | 400 | 500 | 900 | - | 120 |
| 10. | Brook Land | - | - | - | - | 1000 | 300 | 200 | - | - | 1000 | 300 | 200 | - | - | 100 |
| 11. | Shergarhi | 1137 | 618 | - | - | 1000 | 500 | 500 | 500 | - | 500 | 500 | 500 | 500 | - | 130 |
| 12. | Barlowganj | 221 | - | 195 | - | 400 | 200 | 400 | - | - | 300 | 200 | 400 | - | - | 60 |
| 13. | Kincraig | - | - | - | - | 500 | 500 | 700 | - | - | 500 | 500 | 700 | - | - | 110 |
| 14. | Bhattafall | - | - | - | - | 800 | - | - | - | - | 800 | - | - | - | - | 50 |
| 15. | Shishu Niketan | - | 1000 | - | - | 500 | 500 | 700 | - | - | 500 | 500 | 700 | - | - | 110 |
| 16. | Library | - | - | - | - | 500 | 500 | 500 | - | - | 500 | 500 | 500 | - | - | 100 |
| 17. | Vincent Hills | - | - | - | - | 600 | 600 | - | - | - | 600 | 600 | - | - | - | 80 |
| | | | | | | | | | | | 11700 | 10000 | 5900 | 4900 | 500 | 2185 Nos. |
| | | | | | | | | SAY | | | 12000 | 10000 | 6000 | 5000 | 500 | 2000 Nos. |

*ANNEXURE-2*

# Waste Management in Musssoorie : A Case Study

## Introduction

The purpose of this case study is to assess the waste disposal and management facilities in Mussoorie and to see if the existing population pressure has not already overstretched the limits of these facilities. The study also attempts to project the waste management scenario in the near future on the basis of existing trends and expected developments.

Mussoorie Township was planned in the days of the British. Garbage disposal systems put in place at the very beginning are continuing to deliver till date. Growing consumerism and increasing use of plastics is giving way to a 'throw-away' culture, which is aggravating the problem of effective waste disposal generating more waste with an increasing proportion of non-biodegradable garbage.

Waste generated in any township can be classified into three broad categories namely, garbage (solid waste), sewage and gaseous emissions and garbage further classified into biodegradable and non-biodegradable. With the advent of plastics and its use in packaging, a substantial portion of garbage is that which does not degrade with time, or break down into smaller and simpler constituents. This portion when introduced into the environment remains there unchanged in form, for years to come. Non-biodegradable waste (e.g. plastic, films, celluloid batteries, chemicals, glass and other synthetics), is called dry waste in common parlance. On the other hand, wet waste (e.g. paper, plant and animal products) degrades readily under favourable conditions of moisture and temperature. Degraded waste is good quality manure and can be mixed with soil. Thus when left on its own, dry waste does not get recycled into some useful form, whereas wet waste gets recycled into manure. Hence while the load of the latter gets reduced by decomposition, reuse and recycling into an active human ecosystem, that of the former keeps increasing and accumulating as pollutants.

Waste in Mussoorie is generated by its local resident population, visiting tourists, commercial establishments like hotels, holiday resorts, shops and hawkers, and by schools and other institutions. Part of the solid waste is discarded, by those who generate it, down the *khud* or in dirty heaps in the locality. The Municipality sweepers collect the other part in the trailer bins (put in place by the City Board). On account of the toxic emissions of burning waste, the earlier practice of incinerating the collected waste has been discontinued, following protests. Presently the trailer bins are towed to the Gari Khana site where they are tilted to allow the litter to slide down the hillside. Some of the litter slides into the Bhilaru stream, thereby contaminating a precious source of water.

Any effective method of waste management necessitates separation of the dry waste from the wet. The former can then be sold for recycling while the latter is composted to produce manure, which can be sold. Since waste of a township collected at a single place for disposal is in large quantities, segregation at that stage is much too arduous and messy, making it practically impossible. Hence, segregation to be feasible on a wide scale, has to be encouraged at the point of generation, i.e. in kitchens, home or at the restaurant. Waste discarded in bins along the street by people needs to be separated in two different bins for dry and wet waste which can then be collected separately and managed separately.

Of late concerned citizens and organisations have become active in creating awareness amongst the public and exerting pressure on authorities for better management of waste. Besides creating an understanding on the menace caused by waste, they are persuading people to separate wet from dry waste in order to make both types more manageable. They are also involved, though to a limited extent, in actual separation and recycling of dry waste.

## Environmental Implications of Waste

A growing mound of waste is not only an eyesore, it is also a hazard for society. Waste is smelly and a good breeding ground for disease causing bacteria and insects like mosquitoes. In comparison to wet waste, dry waste poses greater problems. Because it does not decompose, its quantity in the environment keeps building up. In addition, plastics do not allow water to penetrate through them. Hence, as in the case of Mussoorie, polythene bags thrown down the *khud*, over time, cover the roots of trees lining the mountain slope, thereby preventing water from reaching the roots. The roots gradually die and the tree gets uprooted. Without trees and their roots to bind mountain soil, the slope becomes prone to landslides. Wet waste too does not decompose when it is tied up in polythene bags for the purpose of disposal. Its plastic packaging blocks water and air needed for its decay. Part of the waste thrown down the slope, slides down to mix into the water streams thereby polluting them.

Already there are reports of average temperatures in Mussoorie rising. Fans, hitherto unseen in Mussoorie, are becoming a regular feature. During one summer season the maximum value of temperature actually touched 39 degree celsius. Natural streams are drying up and rainfall and humidity levels are falling. Deforestation has led to more frequent landslides and just as in any town in the plains, mosquitoes are becoming a menace.

## Methodology for Assessing Status of Waste Management

For a comprehensive assessment of the waste management status in Mussoorie, the need for consultation with every concerned party was appreciated. This included those that generate waste and those that are involved in its disposal. Under both categories attempts were made to include institutions and individuals, commercial and non-commercial establishments, government departments and non-governmental organisations. In order to accomplish the above, meetings were held, interviews were conducted, and a survey was organised for collection of primary information. Over and above this, assessment also relied on general observations, study of documents and reference to publications relating to waste management in general and of Mussoorie in particular.

Notable amongst those with whom meetings were held are the *Mussoorie Nagar Palika,* Health Department, *Jal Nigam,* MDDA, Tourism Department and NGOs like AAPKA and CLEAN.

A hotel schedule was designed (Appendix 2). A sample of approximately 10 per cent of the total hotels and holiday resorts were selected and the owners/managers interviewed. The selection was made so as to account for variations in size of hotel accommodation and location. An institution schedule was also designed. Concerned personnel in a sample of schools, prominent institutions like the Lal Bahadur Shastri National Academy of Administration, Indo-Tibetan Border Police, Tibetan Homes, Dharamshalas, Horse Lines, Archaeological Survey of India, Hospitals, etc., were interviewed.

For interviewing households, a household schedule was designed. Selection of households was based on systematic random sampling method and done from the latest voters list available with the City Board. Wherever the selected voter was not accessible due to death, transfer, temporary absence or error in records, the next name was selected in place of the former. Voters that were selected and lived in very far-off areas were excluded from the survey as their impact is localised and of little consequence to the main city. In this way 226 households were interviewed.

Floating population figures were obtained from the Tourism Department where the estimate is made on the basis of the vehicles crossing the Toll Tax collection point. Average length of stay of a tourist in Mussoorie was also taken on the basis of an estimate arrived at by the Tourism Department.

For estimation of the average quantity of waste generated per capita per day by residents and tourists in Mussoorie and break-up analysis of the waste, results of a survey conducted by an NGO have been used (Appendix-6).

## Quantity of Waste Generated in Mussoorie

Mussoorie's population is divided into the permanent and the floating. While the residents of the town, boarding students in schools, permanent staff in hotels and dharamshalas etc. come in the former category, tourists and temporary staff at commercial establishments that move to Mussoorie only during peak tourist seasons may be included in the latter.

Consumption patterns of these categories differ. Tourists, as compared to residents, tend to generate more waste, as concluded from the survey mentioned earlier (see Appendix 6). This is because, besides regular meals they have a tendency to buy readymade food items off the shelf for consumption. In such items, the weight of the packaging is considerable as compared to the food item itself. Also there is no reuse of the packaging. Fresh litter is generated each time. Heavy waste items like empty glass bottles are discarded by tourists in greater quantities.

## Current Status of Waste Management

Responsibility for management of waste in Mussoorie falls within the purview of the elected City Board, the Health Department and the Water Works Department. This arrangement is similar to that in other cities. In Mussoorie, however, the work of NGOs, institutions and contribution of individual citizens is supplementing the work of government departments. Details of areas covered, sources of funds, facilities available and procedure followed by the City Board, concerned NGOs and public institutions are as follows:

### *City Board*

City Board staff, by its own admission, is unable to access certain areas already enumerated. Even in the wards where they claim to provide services and from where they collect the Conservancy Tax, the Municipality sweeper either does not visit or visits irregularly. As a result households have employed private sweepers or housewives themselves empty out the small dustbins placed outside their premises (into which the household waste is discarded), into the nearest yellow trailer bin or into a garbage heap on the road or over the hillside.

Regular posts of sweepers have been sanctioned to the City Board. This number was fixed at the time of independence when the city's population was less than 10,000. No addition to this number has been made till date despite population levels having touched 40,000. During peak tourist seasons, the Board tries to cope with the load by employing 50 more sweepers on an ad-hoc basis.

Sweepers are provided with one basket each, which they carry on their backs. If the waste collected is wet, it drips onto them as they carry it on their person. Since the yellow bins into which they are supposed to dispose their load is positioned above the level at which they have collected the waste, they are reluctant to carry the burden. Some times they choose the easier option of simply sweeping the garbage down the *khud* or into dirty heaps in the corners.

Even if segregation takes place at source, because the sweeper is carrying a single basket, he/she mixes both the dry and the wet waste in the basket rendering the former exercise useless.

Location of yellow bins is sparse. If more bins are installed, more households would have easier access to them encouraging disposal of garbage in the bins.

Separate bins (blue and green in colour) have been placed along the Mall Road in order to encourage tourists to discard wet and dry wastes separately. However, information about the use of different bins is lacking. Also the design of the bin is not handy. The lids of the covered bins are very heavy discouraging their use.

Collection of dry waste is more attractive than collecting wet waste because the former can be sold. As a result, the Municipality sweeper at times is tempted to selectively collect only the former leaving the latter where he/she finds it.

The single dumping site at Gari Khana cannot take the collective load of the entire city. Waste thrown over the hillside at this site, slides down into the Bhilaru stream contaminating a precious source of water. Dry waste thrown down the Gari Khana hillside has become a permanent dirty feature because left on its own it does not decompose. Municipality and private sweepers, citizens and tourists alike, tend to sweep or throw waste down the *khuds,* because of a lack of awareness or concern. This has left several hillsides dirty.

All areas of Mussoorie that fall within the Municipality limits are supposed to be serviced by the City Board. Even within these limits there are areas where the Board is unable to provide services. This may be on account of difficult terrain which makes access difficult, sparse habitations making them low priority areas, limited staff, equipment and funds. These areas, as per the City Board, are: (1) Dunhill, Bullet, Pan, Chaskhet and Clouds End; (2) George Estate; (3) Hathipaon; (4) Srinagar Estate; (5) Bhilaru; (6) Makdait Village; (7) Jhalki; (8) Suwakholi; (9) Bataghat; (10) Bansagad; (11) Spring Road; and (12) Circular Road.

With a view towards keeping the city clean, the remaining areas have been divided into three wards. *Nayaks* have been appointed for the purpose of monitoring and supervision of the cleaning and disposal activities. The wards are:

1. Jharipani, Barlowganj, Masonic Lodge Landour uptill Picture Palace, Kolukhet, Bhatta Village, etc.
2. From Picture Palace until Basu Cinema and adjoining areas.
3. From Vasu Cinema through Camel's Back Road until Nirankari Bhawan, Motilal Nehru Road and adjoining areas, Lal Bahadur Shastri National Academy of Administration area, Company Bagh and adjoining areas.

A hundred and fifty one posts of sweepers are sanctioned to the City Board, the number having remained constant since the time of independence. Of these sanctioned posts, 143 are filled and the remaining are lying vacant on account of retirement or death of employees. During the peak tourist season when work - load goes up several times, 50 additional sweepers are employed on a contract basis.

The Board owns 19 trailer bins and waste trolleys with capacities of 5,000, 9,000 and 11,000 litres. Of the trailer bins, three bins are out of operation, two of which have been rendered unusable by accidents and the third is undergoing repairs. The remaining 16 trailer bins and waste trolleys are placed all over the city. Besides these, several small dustbins (some coloured green and blue to indicate whether dry or wet waste is to be thrown into them) have been placed all over the city, outside homes, shops, hotels, schools etc. and along the streets. Two tractor trolleys and two tractor trailers are owned by the Board. Along with drivers they are pressed into service for towing the trailer bins and waste trolleys full of waste, from their location, to the dumping site at Gari Khana.

The City Board collects a Conservancy Tax from the establishments where it provides services. The Municipality sweepers are divided among the conservancy staff and the Roadside staff. The former goes from house to house, hotels, schools, etc. and collect the waste from the small dustbins placed outside the premises into the basket carried by him. This load he empties into the nearest yellow trailer bin. The latter collects waste from the dustbins placed along the roads, into the basket carried by him. He also sweeps the streets, collecting the rubbish in the basket. The basket is then emptied into the nearest yellow bin. Wherever waste has been segregated into dry and wet waste at source, i.e., by those who have generated it, the sweeper is supposed to collect the two wastes separately. This is rarely done in practice because the sweeper has only one basket to collect the waste.

Earlier the designated sites were four in number where the municipality had installed incinerators (Gari Khana, below South Road Landour, South of Butcher Khana and Barlowganj). However, this practice has been discontinued, due to the environmental pollution caused by the dioxins released during burning of waste. Presently, for want of a better alternative, the entire garbage collected by the City Board staff is emptied at a single site, i.e., down the Gari Khana hillside forming a landslide of waste. That part of the waste that is biodegradable degrades in the natural course of time. The part that does not degrade, notably plastics, lies there as a permanent feature. With rains and fresh loads of waste pouring in some of the waste slides down along the hillside into the Bhilaru stream.

In order to prevent this downslide and to provide pits where the waste can collect and compost, terraces were constructed along the Gari Khana hillside. Waste thrown from above would slide down and collect into the pits formed by the terraces and decompose there. Due to poor quality of construction the terraces have got washed away with the rains.

**Non-Governmental Organiations**

AAPKA or the Self-Help Environment Programme (SHEP) works in association with the Lal Bahadur Shastri National Academy of Administration. The main thrust of the organisation has been towards effective garbage management both at the individual and the community level. The organisation has tried to create awareness regarding the concept and need for waste management and the concept of separating the biodegradable portion of waste (wet waste) from the non-biodegradable portion (dry waste), and disposing the two portions independently. Attempts were also made to create awareness regarding the harmful effects of use of polythene and at the same time looking for alternatives to polythene. It has also explored alternative ways of reusing polythene bags. Wherever and whenever possible the organisation has tried to draw the attention of the authorities to areas of neglect.

AAPKA's operations, particularly awareness creation, are spread throughout the city. However, the main thrust of its activities has been along the Charleville Road until Happy Valley, along the Mall Road, through Kulri Bazaar until Picture Palace and in the Woodstock School campus.

The main source of funds for AAPKA is from the sale of dry waste. Constituents of dry waste like polythene bags, paper, glass items, tins, batteries have a ready market in Dehradun. The above amount is supplemented by fixed and voluntary contributions from households. Other voluntary contributions add to the funds.

AAPKA has organised some rag pickers under its wings. They

have been equipped with basic equipment like brooms, baskets, gumboots and sticks. Some have also been provided with pushcarts. Besides the above, it has issued Identity Cards to its rag pickers thereby allowing them easier and greater access to homes and hotels. It also provides them with medical facilities and arranges transport for them to Dehradun for selling the dry waste collected by them. Rag pickers, the working hands of this organisation, are not readily accepted by the households, hotels, etc. because of their reputation for stealing. AAPKA rag pickers often run into conflict with Municipality sweepers. Rag pickers selectively collect dry waste, which can be sold. Municipality sweepers thus feel that instead of assisting them in their work, rag pickers trespass in their areas of operation.

For the purpose of creating awareness regarding the concept and need for waste management, catchy and comprehensible slogans like "SAY NO TO PLASTICS", "DO NOT THROW AWAY PLASTIC PACKINGS OR BAGS BUT COLLECT THEM SEPARATELY", "SAVE MUSSOORIE—YOU CAN DO IT" and many others have been coined and communicated using street campaigns, talks, processions, posters, leaflets, stickers, etc. Talks have been delivered in schools and school children involved in distribution of propaganda material. Sanitation Day was organised in April 1995. Households, institutions and commercial establishments have been persuaded to segregate waste generated by them. This involves collecting non-biodegradable waste separately and making it available to the local sweeper or AAPKA rag pickers in separate bags or containers.

AAPKA rag pickers, equipped with basic training and equipment, are sent door to door to collect the separated dry waste. Wherever segregation has not been done already, and it is possible to do it, given the co-operation of the concerned people, the rag pickers are persuaded to segregate the waste themselves. The dry waste so collected is further segregated into polythene bags, glass items, paper, batteries, tin, etc. Stuffed into bags it is carried down to Dehradun for sale to the *kabaris* there. Proceeds from the sale of waste go directly to the rag picker, who has collected it. No share is retained by the organisation.

Training workshops have been conducted for developing and popularising alternative uses of disposed polythene bags and paper. This involved making bags out of discarded polythene bags and paper. This is a cheaper and less energy intensive way of reusing the above waste. Subsequent to this use, they can be sold for recycling into fresh paper and plastic. Efforts have been directed at organising a paper bag making cottage industry.

Companies like AMUL have been persuaded into sponsoring posters and advertising slogans. Others local industrial units like the U.P. Mineral Development Corporation have sponsored dustbins placed along the streets. Effective waste management strategies have been successfully operationalised at Hampton Court School and Woodstock School. All residents of the school, the staff, their resident families and the students have been sensitised and trained to discard dry and wet waste into different bins put in place for this purpose. Polythene bags have been hung in classes into which the students discard wrappings, pencil shavings etc. Dry waste is made available to the rag pickers. For the wet waste pits have been dug on the school premises. The number and size of the pits is determined by the capacity required, the availability of land and depth to which it is possible to dig without encountering any hard obstruction like rocks. Wet waste is thrown into one of the pits. In order to prevent bad smell and breeding of insects, every day or two, ash made available from the school kitchen is thrown in the pit forming a layer over the waste. Every few days, soil is also layered into the pit. When this pit is full, it is covered well with soil and left to decompose. It takes three to six months depending on the season before decomposition is complete and manure is ready. In Hampton Court school, such a system of cycling of pits is managed between seven pits, and school children help when needed. Nobody from outside the school, except the rag pickers, is involved. The school does not pay Conservancy Tax.

CLEAN (Careful Learning from the Environment and Nature) is a self-sustained programme that has directed its efforts in the Landour Raj Mandi Ward and the Cantonment by working with residents, schools and guests houses to separate dry waste from wet waste at the source.

The organisation provides service to Woodstock School, ITM and private residents in the Cantonment. During 1998 many parts of the Jabarkhet Jhalki Ward are expected to be included in an educational programme.

During 1997, 2 tonnes of plastic, primarily polythene bags and 8 tonnes of paper products were recycled through a *Kabari* in Mussoorie. In addition CLEAN provides some special services.

One manager and six trained workers form the basic staff. The staff strength increased by one by the end of 1997 and it is expected that an educational extension programme will develop. A trial cleaning programme is underway at Kempty Falls, with the hope that this will become a self-sustaining project.

CLEAN has encouraged all houses, institutions and guest houses in its area of operation to separate dry waste from wet waste. For management of the wet waste it suggested digging pits on available land in which to compost the organic material. If needed it supplies manpower for digging the pits. Some mixed waste is still thrown down *khuds*. The workers of CLEAN regularly clear the *khuds*. The success of the source separation is shown in the first quarter report for 1998 when 200 kg. less of polythene was collected. In the previous year a great deal was still coming from the *khuds*.

CLEAN workers collect and sort the dry waste before it is sent to a Mussoorie *kabari*. The waste is separated into paper, cardboard, polythene, glass items including broken glass, batteries, tins, etc. Detailed records have been kept for more than three years.

CLEAN has attracted wide interest. Representatives have participated in workshops held by the Environics Academy in Dehradun as well as a one-day workshop in Shimla. A web page is being designed for the Internet. Over 2,000 nylon and cloth carry bags as alternatives to plastic bags have been distributed. Currently CLEAN is working with the Inner Wheel Women's organisation of the Rotary Club. During the monsoon season tree planting is undertaken. Discussions are underway for the setting up of a handmade paper unit following the Solan model in Himchal Pradesh.

**Waste Management by Institutions in Mussoorie**

Apart from households and hotels, other institutions and establishments in Mussoorie were also surveyed to study the solid waste disposal method adopted by them. These methods ranged between self reliance to complete dependence on the Municipality. In the case of some establishments, it was observed that waste is disposed over the hillside or into open pits. This was observed to be on account of either indifference or helplessness due to absence of better alternatives. On the other hand, in some institutions, the waste management set-up was observed to be worth emulating.

*Schools*

Hampton Court School and Woodstock School are completely self-reliant in the management of the waste generated on their premises by the students, staff and their families. For this, those generating the waste themselves have been motivated to participate fully. The method adopted by them involves separation of dry and wet waste at the point of generation itself, i.e. in the kitchen, staff quarters and classrooms. Dry waste is made available to the rag-pickers who sell it. Wet waste is dumped into deep pits dug in the school compound. From time to time ash from the kitchen and soil is layered into the pits. When the pit is full, it is covered with soil and left to decompose. Depending on the weather, it takes three to six months for decomposition to be completed and manure to form. This manure is used in the garden in the school. In the Central School for Tibetans, segregation is partial. Waste disposal and management is undertaken by the residents of the school themselves and a few private sweepers. Some of the vegetable waste is made available to pig-breeders and some dumped into pits in the school. The rest is dumped into the yellow bin close to the school premises.

*Hospitals*

The Community Health Centre (CHC) in Mussoorie utilises the services of the Municipality sweeper. The hospital staff separates out biological waste and bandages from the remaining waste but the sweeper clubs it all together and throws it into an open and untreated pit close by the quarters of some of the hospital staff. The sweeper does, however, pull out the clean polythenes from the waste which he carries away to sell. At the Veterinary Hospital, all the waste collected from the hospital by its sweepers is thrown down the Gari Khana hillside which is close to the hospital compound.

*Restaurants*

Restaurants employ the services of either private or municipality sweepers to remove waste from their premises. They are at best segregating waste only partially. Some are not segregating at all. The restaurants on the Mall Road do not complain of illegal dumping spots close to their premises. However, those in the Kulri Bazaar and at Picture Palace complain about them during the season. Hydraulic bins lie close to the premises of most hotels, i.e. within a distance of 100 meters. The restaurants serviced by the municipality sweeper are satisfied with his service.

*Mazdoor Niwas*

The residents of the Chhapan Kote Mazdoor Niwas, below the Library Bus Stand, discard their waste generated by them over the hillside. A hydraulic bin is positioned within 100 meters of the Niwas but the walk to it is uphill. No municipal sweeper visits the Niwas.

*Dharamshalas*

Laxmi Narayan Dharamshala is collecting the total waste generated by its residents and burning it, despite the presence of a hydraulic bin within a distance of 50 meters. The walk to the bin is, however, uphill and the dharamshala is not utilising the services of municipality sweepers. The staff in the dharamshala is indifferent towards segregating waste.

*Tibetan Homes*

These homes are partially segregating waste. The residents of the homes themselves burn a part of the wet waste and the municipality sweeper carries a part to the nearest yellow bin. The dry waste is carried away by the sweeper for sale. The homes complained about the presence of dirty heaps of waste near their premises.

*Indo-Tibetan Border Police Academy*

The academy manages the waste by pressing its own team of sweepers into service. There is complete segregation of waste with the wet portion being burned and the dry portion being burned in pits.

*Survey of India, Landour*

In this campus, sweepers employed by the Department who are responsible for cleaning the office area and the estate manage the waste. Staff quarters on the estate dispose their waste themselves by throwing it over the *khud.*

*Horse Lines*

In the Horse Line in the LBS National Academy of Administration, horse waste is collected in burrows and wheeled to a *nallah* within the Lines where it is dumped and left to dry up. Since this waste has a very high content of toxins, trees in this *nallah* have been drying up. At the horse stand on the Mall Road, the Municipality Sweeper sweeps horse waste that collects during the day into the hydraulic bin. This bin is positioned right where the horses stand. Sometimes the horses eat garbage from out of this bin itself. Some horses have been reported to die as a result. During the night, the owners of the horses used for commercial purpose tie their horses wherever they can find a place for them. They sweep horse waste that collects through the night over the hillside in the morning.

*APPENDIX - I*

# CIVIC SERVICES AND INFRASTRUCTURE QUESTIONNAIRE

**QUESTION:**

1. Sl. No.

2. Name & Address

3. Educational Level
   (a) Illiterate (b) From Class IInd to Class Vth
   (c) From Class VIth to Class XIIth (d) Higher Education

4. Occupation

5. Status of the Family (Number of Family Members)

6. Income Level (Average Monthly Income of Family)
   (a) Less than Rs. 2000 (b) Between Rs. 2000 to Rs. 5000
   (c) More than Rs. 5000

**Litter and Public Hygiene**

To what extent do you think that the city is littered?

1. Extremely
2. Somewhat
3. Not at all

What causes the maximum litter? Prioritise 1 to 6

Plastic bags

Plastic containers
Bottles
Tins
Kitchen waste
Animal droppings

Who is more responsible for the litter?

1. Tourists
2. Local population

How effective are the safai karamcharis?

1. Very effective
2. Effective
3. Ineffective

Do you think that there are sufficient public bins for litter?
Yes No

Do you think that open hydraulic bins are disturbed by cattle and monkeys? Yes No

Do you think that sufficient hygiene condition is maintained by sellers of eatables? Yes No

Do these sellers dispose off the left-overs and empty/used disposable plates etc. in the public bins? Yes No

Is there satisfactory incineration facility for garbage?
Yes No

Do the drains get choked during the rainy season?
Yes No

Where in your locality do the drains get choked?

Where in the city is there maximum litter?

Where in your locality are public bins/hydraulic bins needed?

What is the best solution for tackling litter?

**Pollution**

Do you think the air is polluted during the peak season?
Yes No

Do you think that the air is polluted all round the year?
Yes No

Do you think that there is noise pollution during the peak season?
Yes No

Do you think that there is noise pollution all round the year?
Yes No

Do you think that there are effective laws for preventing air and noise pollution? Yes No

Where in the town is there maximum air pollution?

Where in the town there is maximum noise pollution?

**Public Conveniences**

Are there sufficient public urinals in the town?
Yes No

Are there sufficient public latrines in the town?
Yes No

Are the urinals and latrines clean? Yes No

What can be done to improve the cleanliness of urinals/latrines?

Is Mussoorie public convenience (urinals/toilets) have water facilities? Yes No

Is there any necessity in Mussoorie to have public bathing place?
Yes No

Are there sufficient public drinking water facilities in the town?
Yes No

Are there sufficient rain shelters in the town?
Yes No

Are there sufficient benches for sitting in public places?
Yes No

Where in your locality should urinals/latrines be located?

Where in your locality should rain shelter and benches be located?

Where in your locality should public drinking water facility be provided?

**Roads**

Is maintenance of public roads good? Yes No

Is the road network in the town adequate? Yes No

Is the width of the roads sufficient? Yes No

Do you think that the roads get overcrowded during the season?
Yes No

Do you feel like walking on the Mall Road during peak hours of the season Yes No

Is there sufficient parking space? Yes No

Have you experienced open manhole in the road which is prone to an accident? Yes No

Where in the town is there maximum congestion?

Where in the town should roads be widened?

Where in your locality can parking space be created?

**Water and Electricity**

Are you satisfied with the quality of drinking water?
Yes No

Are you satisfied with the supply/availability of water during season? Yes No

Are you satisfied with the supply/availability of water all round the year? Yes No

Is there proper management for natural water sources maintenance in Mussoorie? Yes No

Are you satisfied with the supply of electricity during season? Yes No

Are you satisfied with the supply of electricity all round the year?
Yes No

Do power cuts adversely affect your work and productivity? (During season or all around the year) Yes No

If yes when? During season
All round the year

Are you satisfied with the maintenance of the electricity department? Yes No

Do you think that the electricity department is understaffed?
Yes No

Do you think that there are sufficient street lights in the town?
Yes No

Which is the worst lit area in your locality?

**Supplies**

Do you think that there is a general shortage of milk, vegetables, fruits, eggs and other essential supplies during the season?
Yes No

Do you think that the prices of such commodities become unbearable during the season? Yes No

Do you have to wait long for purchasing household items during peak hours of the season? Yes No

Where should new markets be located?
Did you observe/experience anytime that harassment and exploitation of the tourist and general public by hotel-owners/shop keepers/rickshaw pullers/photographers, during the season?
Yes No

**Construction Activities & Accommodation**

Do you think that the construction activities in the town are planned satisfactorily? Yes No

Do you think that construction activities in the town have caused congestion and inconvenience? Yes No

Do you think that the natural beauty and scenic views are obstructed by the construction activity in the town? Yes No

Do you think that the building laws are being implemented?
Yes No

Where in the town has there been the most unplanned growth of buildings and population?

Is it right that during season because of insufficient accommodation tourist are forced to sleep on the Mall road?

Yes No

Did you or your friends experience during season any difficulties in finding accommodation in Mussoorie? Yes No

If yes, what is the solution of this problem in your opinion?

**Transport**

Do you experience traffic jams? (During season or all around the year)

1. Very frequently
2. Frequently
3. Sometimes
4. Rarely

Do you think that the traffic management is effective during season?

Yes No

Do you think that the traffic management is effective all round the year? Yes No

Do you think picnic spots like Kempty are easily accessible by bus/taxi? (During season or all around the year)

Yes No

Are there adequate public transport facilities? (During season or all around the year) Yes No

Do you have to wait in long queues for long period for travel by public transport to and from Dehradun during season?

Yes No

Are the buses and taxis overcrowded? (During season or all around the year) Yes No

Are the bus stands overcrowded? (During season or all around the year) Yes No

Which spot in your locality experiences the maximum traffic jams? (During season or all around the year)

**Entertainment**

Do you think that there are sufficient entertainment facilities in the town? Yes No

Are there sufficient cultural activities in the town?
Yes No

Are there sufficient sports facilities in the town?
Yes No

Which new facilities should be created and where?

Would you like having a botanical garden in Mussoorie?

Which tourist places should be developed?

**Health**

Are there sufficient hospital facilities in the town?
Yes No

Do you think there are sufficient medical specialists in town?
Yes No

Which area in town is worst covered by medical facilities?

Is the food served in the hotels (restaurants) having satisfactory quality standard? Yes No

**Education**

Do your children get admission easily in schools?
Yes No

Are schools overcrowded? Yes No

Are there sufficient buildings/rooms in schools?
Yes No

Is there adequate provision for higher education in Mussoorie?
Yes No

**Animals**

Do you think that there are too many stray dogs, cattle, pigs etc?
Yes No

Are there sufficient facilities for the disposal of dead animals?
Yes No

Do you think that there are satisfactory regulations for pets?
Yes No

Are there sufficient veterinary facilities for animals?
Yes No

Which area in town is too far from veterinary facilities?

**Others**

Are the facilities of cremation grounds, cemetry satisfactory?
Yes No

Where do you think additional cremation and burial facilities can be created?

Do you think that the infrastructure facilities are sufficient to cater to the population of Mussoorie during the season?

Yes No

Is there any particular place for disposing the debris?

Which infrastructural facility in your opinion needs the highest priority?

How often has your journey to and from Dehradun been obstructed by landslides?

1. Very often
2. Sometimes
3. Rarely

At which locations have you experienced landslides?

In Mussoorie are there enough opportunities for employment? (professional work)

If not, in your opinion which employment (professional work) should be encouraged?

**Any other observations/suggestions.**

*APPENDIX - II*

## QUESTIONNAIRE—WASTE MANAGEMENT

1. Sl. No.:

2. Name & Address:

3. Location: Top of Hill    Slope    Bottom of Valley

4. Education level:
   Head of H. H.    Spouse
   (1: Uneducated, 2: till class 5, 3: till class 12, 4: higher)

5. Occupation:

6. Family size

7. Kitchen Expenditure

8. Income group: (Family income per year)
   Less then Rs. 10,000
   Between Rs. 10,000 and Rs. 50,000
   Between Rs. 50,000 and Rs. 1,00,000
   More than Rs. 1,00,000

9. How is disposal of waste done:
   Municipality sweeper    Private sweeper
   Self

10. Is it segregated    Yes    No

11. If yes, is segregated waste sold Yes No

12. How far is the hydraulic bin
13. Location of hydraulic bin : Above level Below level

14. Is clearing of bin regular: Yes No
15. Is sweeper regular: Yes No

16. Is sweeper effective: Yes No

17. Is there an illegal dumping spot close to the premises
    Yes No

18. Are you prepared to segregate waste into wet and dry waste?
    Enthusiastic Willing
    Indifferent Reluctant

19. Nature of sewerage system?
    Sewerage system Soak-pit/septic tank/other

20. Any problem faced with it?

**Hotel Questionnaire**

1. Name :

2. Capacity :

3. Address :

4. Location : Top of Hill Slope Bottom of Valley

5. Star category :

6. How much waste :

7. How is the disposal of waste done :
   Municipality sweeper    Private sweeper    Self

8. Is it segregated    Yes    No    Partially

9. If yes, is segregated waste sold    Yes    No

10. How far is the hydraulic bin

11. Location of Hydraulic bin
    Above level    Below level

12. Is clearing of bin regular :    Yes    No

13. Is sweeper regular :    Yes    No

14. Is sweeper effective :    Yes    No

15. Is there an illegal dumping spot close to the premises:
    Yes    No

16. Are you prepared to segregate waste into wet and dry waste?
    Enthusiastic    Willing    Indifferent    Reluctant

17. Nature of sewerage system?
    Sewerage system    Soak-pit / septic tank/ other

18. Any problem faced with it ?

**Institution Questionnaire**

1. Name :

2. Capacity :

3. Address :

4. Location : Top of Hill Slope Bottom of Valley

5. How much waste :

6. How disposal of waste is done :
Municipality sweeper Private sweeper Self

7. Is it segregated Yes No Partially

8. If yes, is segregated waste sold Yes No

9. How far is the hydraulic bin

10. Location of Hydraulic bin Above level Below level

11. Is clearing of bin regular : Yes No

12. Is sweeper regular : Yes No

13. Is sweeper effective : Yes No

14. Is there an illegal dumping spot close to the premises:
Yes No

15. Are you prepared to segregate waste into wet and dry waste?
Enthusiastic Willing Indifferent Reluctant

16. Nature of sewerage system?
Sewerage system Soak-pit / septic tank/ other

17. Any problem faced with it?

*APPENDIX - III*

## CIVIC SERVICES AND INFRASTRUCTURE

## Distribution of Respondents by City Wards

| Sl. No. | Ward No. | Name of the Ward | Number of Residents' responses |
|---|---|---|---|
| 1. | 1 | Allen Wyne Burg | 10 |
| 2. | 2 | Jharipani | 15 |
| 3. | 3 | Barllow–gunj | 15 |
| 4. | 4 | Bhilaru | 11 |
| 5. | 5 | Landour–Raj Mandi | 13 |
| 6. | 6 | Charliville | 13 |
| 7. | 7 | Bhad–raj | 14 |
| 8. | 8 | Kachari | 11 |
| 9. | 9 | Hampton Court | 16 |
| 10. | 10 | Mussoorie Club–New Bharat Hotel | 11 |
| 11. | 11 | Hakmans & Camels Back Road | 12 |
| 12. | 12 | Jaber-Khatt & Jhalki | 15 |
| 13. | 13 | Vincent Hill (Gurunanak School) | 13 |
| 14. | 14 | Waverly Convent | 12 |
| 15. | 15 | Oak Grove School | 13 |
| 16. | 16 | Nagar Palika Parishad | 11 |
| 17. | 17 | Polo Ground | 10 |
| 18. | 18 | Library | 15 |
| 19. | 19 | Landour (South) | 13 |
| 20. | 20 | Nala Pani Dhobhi Ghat | 12 |
| 21. | 21 | Landour (North) | 10 |
| 22. | 22 | Bakery Hill (Near City Board) | 14 |
| 23. | 23 | Malacalf | 10 |
| 24. | 24 | Kulri Bazar | 12 |
| 25. | 25 | Happy Valley | 10 |

Government/Quasi-Government organisations - 18
City Board Members—17, Hotliers—13.

*APPENDIX - IV*

# TOURIST INFLOW IN MUSSOORIE

| Months | 1990 | 1991 | 1992 | 1993 | 1994 | 1995 | 1996 |
|---|---|---|---|---|---|---|---|
| January | 57372 | 50000 | 60000 | 48000 | 50000 | 45000 | 50000 |
| | 44 | 15 | 36 | 27 | 34 | 60 | 132 |
| February | 66536 | 60000 | 70000 | 75000 | 70000 | 55000 | 60000 |
| | 64 | 49 | 43 | 38 | 36 | 166 | 138 |
| March | 93536 | 95000 | 100000 | 100000 | 90000 | 75000 | 85000 |
| | 53 | 71 | 75 | 73 | 135 | 308 | 286 |
| April | 126301 | 130333 | 130000 | 132000 | 120000 | 100000 | 125000 |
| | | 50 | 140 | 234 | 127 | 231 | 236 |
| May | 182264 | - | 215000 | 180000 | 160000 | 130000 | 110000 |
| | 75 | 124 | 115 | 132 | 219 | 226 | 238 |
| June | 270263 | - | 325000 | 260000 | 280000 | 250000 | 300000 |
| | 125 | 130 | 109 | 69 | 184 | 256 | 230 |
| July | 158500 | - | 290000 | 280000 | 295000 | 300000 | 320000 |
| | 105 | 138 | 135 | 83 | 154 | 267 | 270 |
| August | 100000 | - | 125000 | 70000 | 60000 | 55000 | 65000 |
| | 140 | 111 | 103 | 68 | 160 | - | 280 |
| September | 80000 | - | 121000 | 78000 | 5000 | 60000 | 65000 |
| | 114 | 175 | 94 | 95 | - | 180 | 220 |
| October | 80000 | - | 155000 | 95000 | 15000 | 80000 | 75000 |
| | 174 | 115 | 137 | 154 | 110 | - | 78 |
| November | 75000 | - | 105000 | 90000 | 35000 | 60000 | 72000 |
| | 51 | 63 | 105 | 104 | 105 | 153 | 136 |
| December | 80000 | - | 60000 | 65000 | 40000 | 70000 | 92000 |
| | 29 | 28 | 32 | 99 | 120 | - | 140 |
| Total | 1369772 | 1500000 | 1756000 | 1473000 | 1220000 | 1280000 | 1419000 |
| | 1041 | 1069 | 1127 | 1176 | 1384 | 1847 | 2423 |

*Note:* The First row represents Indian Tourists and the second represents foreign tourists

(-) represents absence of data.

*APPENDIX - V*

# MONTH-WISE VARIATION IN POPULATION PRESSURE -1997

| Month | Permanent Population | Number of Tourists | Tourist Days @ 3 for May/ June2 for other months | Tourists per day | Population pressure per day |
|---|---|---|---|---|---|
| January | 23930 | 51390 | 102780 | 3315 | 27245 |
| February | 24423 | 65666 | 131332 | 4690 | 29113 |
| March | 40892 | 91361 | 182722 | 5894 | 46786 |
| April | 40892 | 124194 | 248388 | 8280 | 49172 |
| May | 40892 | 164164 | 492492 | 15887 | 56779 |
| June | 40892 | 282648 | 847944 | 28265 | 69157 |
| July | 40892 | 275510 | 551020 | 17775 | 58667 |
| August | 40892 | 79941 | 159882 | 5157 | 46049 |
| September | 40892 | 68521 | 137042 | 4568 | 45460 |
| October | 40892 | 82223 | 164446 | 5434 | 46326 |
| November | 40892 | 71803 | 143606 | 4853 | 45745 |
| December | 25247 | 68521 | 137042 | 4421 | 28844 |

| **In Season Population** | **Winter Population** |
|---|---|
| Total Residents = 34864 | December and February |
| Boarders in Public Schools=4267 | 65% of residents(22662) + Boarders in Tibetan School (1761) =24423 |
| Boarders in Tibetan Schools =1761 | January |
| Total = 40892 | 65% of residents (22662) + Children in SOS Children's Village (1268) = 23930 |

# SCENARIO I

## MONTH-WISE VARIATION IN POPULATION PRESSURE—2001

| Months | Permanent Population | Number of Tourists | Tourist Days @ 3 for May/ June2 for other months | Tourists per day | Population pressure per day |
|---|---|---|---|---|---|
| January | 25788 | 52624 | 105248 | 3395 | 29183 |
| February | 26281 | 67242 | 134484 | 4803 | 31084 |
| March | 43751 | 93553 | 187106 | 6036 | 46787 |
| April | 43751 | 127174 | 254348 | 8478 | 52229 |
| May | 43751 | 164164 | 492492 | 15887 | 59638 |
| June | 43751 | 289431 | 868293 | 28943 | 72694 |
| July | 43751 | 282122 | 564244 | 18201 | 61952 |
| August | 43751 | 79339 | 158678 | 5119 | 48870 |
| September | 43751 | 70165 | 140330 | 4678 | 48429 |
| October | 43751 | 86245 | 172490 | 5564 | 49315 |
| November | 43751 | 74550 | 149100 | 4970 | 48721 |
| December | 26281 | 70165 | 140330 | 4527 | 30808 |

| **In Season Population** | **Winter Population** |
|---|---|
| Total Residents =37723 | December and February |
| Boarders in Public Schools=4267 | 65% of residents(24520) + Boarders in Tibetan School (1761) =26281 |
| Boarders in Tibetan Schools =1761 | January |
| Total = 43751 | 65% of residents (24520) + Children in SOS Children's Village (1268) = 25788 |

# SCENARIO I

## MONTH-WISE VARIATION IN POPULATION PRESSURE—2006

| Months | Permanent Population | Number of Tourists | Tourist Days @ 3 for May/ June2 for other months | Tourists per day | Population pressure per day |
|---|---|---|---|---|---|
| January | 28301 | 54202 | 108404 | 3497 | 31798 |
| February | 28794 | 69259 | 138518 | 4947 | 33741 |
| March | 47618 | 96360 | 192720 | 6217 | 53835 |
| April | 47618 | 130989 | 261978 | 8733 | 56351 |
| May | 47618 | 173147 | 519441 | 16756 | 64374 |
| June | 47618 | 298114 | 894342 | 29811 | 77429 |
| July | 47618 | 290586 | 581172 | 18747 | 66365 |
| August | 47618 | 84315 | 168630 | 5440 | 53358 |
| September | 47618 | 72270 | 144540 | 4815 | 52433 |
| October | 47618 | 88832 | 177664 | 5731 | 53349 |
| November | 47618 | 76787 | 153574 | 5119 | 52737 |
| December | 28794 | 72270 | 144540 | 4663 | 33457 |

| In Season Population | Winter Population |
|---|---|
| Total Residents = 41590<br>Boarders in Public Schools=4267<br>Boarders in Tibetan Schools =1761<br><br>Total = 47618 | <u>December and February</u><br>65% of residents (27033) + Boarders in Tibetan Schools (1761) =28794<br><u>January</u><br>65% of residents (27033) + Children in SOS Children's Village (1268) = 28301 |

# SCENARIO I

## MONTH-WISE VARIATION IN POPULATION PRESSURE -2011

| Months | Permanent Population | Number of Tourists | Tourist Days @ 3 for May/ June2 for other months | Tourists per day | Population pressure per day |
|---|---|---|---|---|---|
| January | 31072 | 55829 | 111658 | 3602 | 34674 |
| February | 31565 | 71383 | 142766 | 5027 | 36592 |
| March | 51881 | 99251 | 198502 | 6403 | 58284 |
| April | 51881 | 134919 | 269838 | 8995 | 60876 |
| May | 51881 | 178341 | 535023 | 17259 | 69140 |
| June | 51881 | 307058 | 921174 | 30706 | 82587 |
| July | 51881 | 299304 | 598608 | 19310 | 71191 |
| August | 51881 | 86844 | 173688 | 5603 | 57484 |
| September | 51881 | 74438 | 148876 | 4962 | 56843 |
| October | 51881 | 71497 | 142994 | 5903 | 57784 |
| November | 51881 | 79091 | 158182 | 5273 | 57154 |
| December | 31565 | 74438 | 148876 | 4802 | 36367 |

| In Season Population | Winter Population |
|---|---|
| Total Residents =45853<br>Boarders in Public Schools=4267<br>Boarders in Tibetan Schools =1761<br><br>Total = 51881 | December and February<br>65% of residents (29804) + Boarders in Tibetan Schools (1761) =31565<br>January<br>65% of residents (29804) + Children in SOS Children's Village (1268) = 31072 |

# SCENARIO I

## MONTH-WISE VARIATION IN POPULATION PRESSURE—2016

| Months | Permanent Population | Number of Tourists | Tourist Days @ 3 for May/ June2 for other months | Tourists per day | Population pressure per day |
|---|---|---|---|---|---|
| January | 34127 | 57503 | 115006 | 3710 | 37837 |
| February | 34620 | 73477 | 146954 | 5067 | 39687 |
| March | 56581 | 102228 | 204456 | 6595 | 63176 |
| April | 56581 | 138967 | 277934 | 9264 | 65845 |
| May | 56581 | 183692 | 551076 | 17777 | 74348 |
| June | 56581 | 316269 | 948807 | 31627 | 88208 |
| July | 56581 | 308283 | 616566 | 19889 | 76470 |
| August | 56581 | 89450 | 178900 | 5771 | 62352 |
| September | 56581 | 76671 | 153342 | 5111 | 61692 |
| October | 56581 | 94242 | 188484 | 6080 | 62661 |
| November | 56581 | 81463 | 162926 | 5431 | 62012 |
| December | 34620 | 76671 | 153342 | 4946 | 39566 |

| In Season Population | Winter Population |
|---|---|
| Total Residents = 50553<br>Boarders in Public Schools=4267<br>Boarders in Tibetan Schools =1761<br><br>Total = 56581 | December and February<br>65% of residents(32859) + Boarders in Tibetan School (1761) =34620<br>January<br>65% of residents (32859) + Children in SOS Children's Village (1268) = 34127 |

# SCENARIO I

## MONTH-WISE VARIATION IN POPULATION PRESSURE—2021

| Months | Permanent Population | Number of Tourists | Tourist Days @ 3 for May/ June2 for other months | Tourists per day | Population pressure per day |
|---|---|---|---|---|---|
| January | 37496 | 59229 | 118458 | 3821 | 41317 |
| February | 37989 | 75681 | 151362 | 5406 | 43395 |
| March | 61763 | 105295 | 210590 | 6793 | 68556 |
| April | 61763 | 143136 | 286272 | 9542 | 71305 |
| May | 61763 | 189203 | 567609 | 18310 | 80073 |
| June | 61763 | 325757 | 977271 | 32576 | 94339 |
| July | 61763 | 317531 | 635062 | 20486 | 82249 |
| August | 61763 | 92133 | 184266 | 5944 | 67707 |
| September | 61763 | 78971 | 157942 | 5265 | 67028 |
| October | 61763 | 97069 | 194138 | 6262 | 68025 |
| November | 61763 | 83907 | 167814 | 5594 | 67357 |
| December | 37989 | 78971 | 157942 | 5095 | 43084 |

| In Season Population | Winter Population |
|---|---|
| Total Residents = 55735<br>Boarders in Public Schools=4267<br>Boarders in Tibetan Schools =1761<br><br>Total = 61763 | December and February<br>65% of residents (36228) + Boarders in Tibetan Schools (1761) =37989<br>January<br>65% of residents (36228) + Children in SOS Children's Village (1268) = 37496 |

# SCENARIO II

## MONTH-WISE VARIATION IN POPULATION PRESSURE—2001

| Months | Permanent Population | Number of Tourists | Tourist Days @ 3 for May/ June2 for other months | Tourists per day | Population pressure per day |
|---|---|---|---|---|---|
| January | 26286 | 52624 | 105248 | 3395 | 29681 |
| February | 26779 | 67242 | 134484 | 4803 | 31582 |
| March | 44518 | 93553 | 187106 | 6036 | 50554 |
| April | 44518 | 127174 | 254348 | 8478 | 52996 |
| May | 44518 | 164164 | 492492 | 15887 | 60405 |
| June | 44518 | 289431 | 868293 | 28943 | 73461 |
| July | 44518 | 282122 | 564244 | 18201 | 62719 |
| August | 44518 | 79339 | 158678 | 5119 | 49637 |
| September | 44518 | 70165 | 140330 | 4678 | 49196 |
| October | 44518 | 86245 | 172490 | 5564 | 50082 |
| November | 44518 | 74550 | 149100 | 4970 | 49488 |
| December | 26779 | 70165 | 140330 | 4527 | 31306 |

| In Season Population | Winter Population |
|---|---|
| Total Residents = 38490<br>Boarders in Public Schools=4267<br>Boarders in Tibetan Schools =1761<br><br>Total = 44518 | December and February<br>65% of residents (25018) + Boarders in Tibetan Schools (1761) =26779<br>January<br>65% of residents (25018) + Children in SOS Children's Village (1268) = 26286 |

# SCENARIO II

## MONTH-WISE VARIATION IN POPULATION PRESSURE—2006

| Months | Permanent Population | Number of Tourists | Tourist Days @ 3 for May/ June2 for other months | Tourists per day | Population pressure per day |
|---|---|---|---|---|---|
| January | 29539 | 54202 | 108404 | 3497 | 33036 |
| February | 30032 | 69259 | 138518 | 4947 | 34979 |
| March | 49522 | 96360 | 192720 | 6217 | 55739 |
| April | 49522 | 130989 | 261978 | 8733 | 58255 |
| May | 49522 | 173147 | 519441 | 16756 | 66278 |
| June | 49522 | 298114 | 894342 | 29811 | 79363 |
| July | 49522 | 290586 | 581172 | 18747 | 68269 |
| August | 49522 | 84315 | 168630 | 5440 | 54962 |
| September | 49522 | 72270 | 144540 | 4815 | 54337 |
| October | 49522 | 88832 | 177664 | 5731 | 55253 |
| November | 49522 | 76787 | 153574 | 5119 | 54641 |
| December | 30032 | 72270 | 144540 | 4663 | 34695 |

| In Season Population | Winter Population |
|---|---|
| Total Residents = 43494<br>Boarders in Public Schools=4267<br>Boarders in Tibetan Schools =1761<br><br>Total = 49522 | December and February<br>65% of residents (28271) + Boarders in Tibetan Schools (1761) =30032<br>January<br>65% of residents (28271) + Children in SOS Children's Village (1268) = 29539 |

# SCENARIO II

## MONTH-WISE VARIATION IN POPULATION PRESSURE—2011

| Months | Permanent Population | Number of Tourists | Tourist Days @ 3 for May/ June2 for other months | Tourists per day | Population pressure per day |
|---|---|---|---|---|---|
| January | 33214 | 55829 | 111658 | 3602 | 36816 |
| February | 33707 | 71383 | 142766 | 5027 | 38734 |
| March | 55176 | 99251 | 198502 | 6403 | 61579 |
| April | 55176 | 134919 | 269838 | 8995 | 64171 |
| May | 55176 | 178341 | 535023 | 17259 | 72435 |
| June | 55176 | 307058 | 921174 | 30706 | 85882 |
| July | 55176 | 299304 | 598608 | 19310 | 74486 |
| August | 55176 | 86844 | 173688 | 5603 | 60779 |
| September | 55176 | 74438 | 148876 | 4962 | 60138 |
| October | 55176 | 91497 | 182994 | 5903 | 61076 |
| November | 55176 | 79091 | 158182 | 5273 | 60449 |
| December | 33707 | 74438 | 148876 | 4802 | 38509 |

| In Season Population | Winter Population |
|---|---|
| Total Residents =49148<br>Boarders in Public Schools=4267<br>Boarders in Tibetan Schools =1761<br><br>Total = 55176 | December and February<br>65% of residents (31946) + Boarders in Tibetan Schools (1761) =33707<br>January<br>65% of residents (31946) + Children in SOS Children's Village (1268) = 33214 |

# SCENARIO II

## MONTH-WISE VARIATION IN POPULATION PRESSURE—2016

| Months | Permanent Population | Number of Tourists | Tourist Days @ 3 for May/ June2 for other months | Tourists per day | Population pressure per day |
|---|---|---|---|---|---|
| January | 37367 | 57503 | 115006 | 3710 | 41077 |
| February | 37860 | 73477 | 146954 | 5067 | 42927 |
| March | 61565 | 102228 | 204456 | 6595 | 68160 |
| April | 61565 | 138967 | 277934 | 9264 | 70829 |
| May | 61565 | 183692 | 551076 | 17777 | 79342 |
| June | 61565 | 316269 | 948807 | 31627 | 93192 |
| July | 61565 | 308283 | 616566 | 19889 | 81454 |
| August | 61565 | 89450 | 178900 | 5771 | 67336 |
| September | 61565 | 76671 | 153342 | 5111 | 66676 |
| October | 61565 | 94242 | 188484 | 6080 | 67645 |
| November | 61565 | 81463 | 162926 | 5431 | 66996 |
| December | 37860 | 76671 | 153342 | 4946 | 42806 |

| In Season Population | Winter Population |
|---|---|
| Total Residents = 55537<br>Boarders in Public Schools=4267<br><br>Boarders in Tibetan Schools =1761<br><br>Total = 61565 | December and February<br>65% of residents(36099) + Boarders in Tibetan School (1761) =37860<br>January<br>65% of residents (36099) + Children in SOS Children's Village (1268) = 37367 |

# SCENARIO II

## MONTH-WISE VARIATION IN POPULATION PRESSURE—2021

| Months | Permanent Population | Number of Tourists | Tourist Days @ 3 for May/ June2 for other months | Tourists per day | Population pressure per day |
|---|---|---|---|---|---|
| January | 42060 | 59229 | 118458 | 3821 | 45881 |
| February | 42553 | 75681 | 151362 | 5406 | 47959 |
| March | 68785 | 105295 | 210590 | 6793 | 75578 |
| April | 68785 | 143136 | 286272 | 9542 | 78327 |
| May | 68785 | 189203 | 567609 | 18310 | 87095 |
| June | 68785 | 325757 | 977271 | 32576 | 101361 |
| July | 68785 | 317531 | 635062 | 20486 | 89271 |
| August | 68785 | 92133 | 184266 | 5944 | 74729 |
| September | 68785 | 78971 | 157942 | 5265 | 74050 |
| October | 68785 | 97069 | 194138 | 6262 | 75047 |
| November | 68785 | 93907 | 187814 | 5594 | 74379 |
| December | 42553 | 78971 | 157942 | 5095 | 47648 |

| **In Season Population** | **Winter Population** |
|---|---|
| Total Residents = 62757 | December and February |
| Boarders in Public Schools=4267 | 65% of residents (40792) + Boarders in Tibetan Schools (1761) =42553 |
| Boarders in Tibetan Schools =1761 | January |
| Total = 68785 | 65% of residents (40792) + Children in SOS Children's Village (1268) = 42060 |

*APPENDIX - VI*

# PER CAPITA WASTE GENERATION BY PERMANENT POPULATION

In order to determine the per capita waste generation and composition of waste generated by the permanent population of Mussoorie, the following study was conducted by an independent organisation.

(1) 25 dustbins were placed at selected locations in the town each day of the one month period between 15th May to 15th June,1997.

(2) Of these, some were placed in households across the town, 5 in hotels (1 in Landour, 2 in Kulri, 1 at the Library Point and 1 on the Charleville Road).

(3) The concerned residents were requested to discard waste in these dustbins alone.

(4) Waste that collected in each of the bins was brought to a central point every day and weighed.

(5) Waste was then segregated into wet and dry portions and each weighed separately.

(6) Dry waste was then separated into broad constituents like paper, glass, polythene and PVC packaging, debris, metal etc. and each constituent weighed separately.

***Results***

1. Per capita waste generated by the permanent population was 350 grams per day.[8]
2. Of this 74 per cent was wet waste and 26 per cent was dry waste.

---

[8] In India, per capita waste generation lies between 0.15-0.50 Kg. per day.

3. The constituents of dry waste were in the following percentage of the total waste:

| Constituent | Per cent of Total Waste |
|---|---|
| Paper | 2 |
| Glass | 6 |
| Plastic/Polythene/PVC packaging | 7 |
| Debris | 4 |
| Metal | 3 |
| Cloth | 2 |
| Wood/Bones | 1 |
| Battery Cells | 1 |

***Per Capita Waste Generation by Floating Population***

1. Dustbins were placed in selected hotels across the town each day of the one month period between 15th May to 15th June,1997. Some dustbins were also placed along the main road at important locations like Rialto Cinema Hall, Library Point and Landour Chowk.
2. Waste disposed by tourists during the night was collected from the dustbins placed in the hotel rooms and weighed daily.
3. Waste generated by the tourists in the hotel during the day was estimated by subtracting the waste generated by the staff of the hotel from the total waste of the hotel.
4. Waste generated by tourists during the day outside the hotel was estimated on a daily basis—weighing the waste collected in the dustbins placed at select points on the main road.
5. Waste collected from the above locations was separated into its wet and dry components and both weighed separately.

***Results***

1. Per capita waste generated by the floating population in Mussoorie was estimated to be 400 grams per day.

2. Of this 70 per cent was wet waste and 30 per cent was dry waste.

*Limitations of the Study*

1. The survey was limited to a period of one month. Consumption and waste generation patterns during that season are being extrapolated to rest of the year.
2. In order to determine the waste generated by tourists in Mussoorie during the day outside the hotels, dustbins were placed in select locations along the Mall Road. The positioning of the dustbins was at prominent tourist points but this still left many areas uncovered.

# INDEX


AAPKA, 105, 106
Aerial photographs, 60
Aglad Water Pumping Scheme, 89, 93
Air pollution, 16, 29
AMUL, 107
Assessment, 53

Basic needs, crrying capacity, 45-46
Basic parameters, 82
Bhadraj, 65
Bhilaru stream, 53
Boarding schools, 37
Build-up area, 64, 65, 68-69
  expension of, 81
  proportion of, 69
Building regulation, 16

Cantonment area, 11
  population of, 36
Careful Learning from Environment and Nature (CLEAN), 100, 107, 108
Cartographic data, 60
City administration, 18-19, 34
City Board, 101
  services, 103
City expansion, limits to, 60-64
City infrastructure, 81
City population, 35
Civic facilities, 24
Civic services, 26-27, 112-26
  and infrastructure, 112-26
Commercial consideration, 82
Community Health Centre (CHC), 109
Conservancy tax, 101, 104, 107
Considerations, 82
Craig Top, 92, 93
Current carrying capacity, 48

Data, 17, 60
Decentralisation, 22
Deforestation, 99
Degraded waste, 98
Dense forest land, 64, 65, 69
Development Authority, 82
*Dharamshalas*, 12, 53, 100, 110
Drinking water sources, 83
Dry waste, 97, 102, 103
  recycling of, 98
Dustbins, 102, 140

Economy, 53
Education, 23, 28-29
Educational achievements, 24-25
Electricity supply, 85
Environmental pollution, 104
  problems, 16

Finance Committee, 18
Floating population, per capita was generation by, 140-41
Forest Conservation Act, 13

Garbage disposal system, 97
Garden Committee, 18
Garhwal Jal Sansthan, 86
Gari Khana, 102, 103
Growing consumerism, 97
*Gurudwaras*, 53

Hampton Court School, 107, 109
Happy Valley, 65
Health Committee, 18
Horse lines, 111
Hospitals, 109
Hospital beds, 45
  projected demand for, 55-57
  requirement of, 58-59
Hotel Owner's Association, 17, 22
Housing, 45
  shortages of, 28
Hydraulic bins, 110

Identity cards, 106
Illegal construction, 93
Indo-Tibetan Border Police Academy, 100, 111
Infrastructure facilities, 23
Intra-city disparity, 30-34

Jabharkhet Jhalki, 65
*Jal Nigam*, 17, 48, 100
*Jal Sansthan*, 18, 25, 85, 87, 88, 89, 91, 92
Jaypee Resorts, 88
Jhari Pani, 65

*Kabaris*, 106, 108
*Khuds*, 98, 99, 102, 103, 108
Kutchery, 65

Lal Bahadur Shastri National Academy of Adminisration (LBSNAA), 100, 105, 111
Land conversion matrix, 80
Land degradation, 29-30
Land development, 28-29
Landour Contonment Board, 17, 30
Landour Raj Mandi ward, 107
Landslids, 29-30, 99
Landuse categories, 60, 61, 77
  distribution, 60

Major indicative solutions offered, 31
*Mazdoor Niwas*, 110
Medical facilities, 83
Monitoring carring capacity, 16
Mosquitoes, 99
Municipal Board, 18
Municipal Board Area, population of, 36
Municipal Ward Boundary Map, 60
Mussoorie, landuse in, 65, 68, 70, 76
  measuring carring capacity, 35-36
Mussoorie-Dehradun Development Authority (MDDA), 12, 100
Mussoorie Municipality, annual income, 20
  annual expenditure, 21
Mussoorie Nagar Palika, 100
Mussoorie Township, 97
Mussoorie Urban Agglomeration, 11, 40

Nagar Palika Act, 18
*Nayaks*, 103
New water pump scheme, 92
Noise pollution, 29
Non-Biodegradable waste, 97
Non-Governmental Organisations (NGOs), 105

Objectives, 13-16
Ongoing works, 89
Open forestland, 64, 69

Parameters, 17, 25
Participatory processes, 82
Perceived indicative solutions, 30
Permanant population, 11, 81
  estimates of, 37-38, 43-44
  per capita waste generation, 139-41
  water demand, 80, 92
Pollution, 29-30, 83
Pollution Control Board, 25
Polythene bags, 99, 105, 107
Population, growth rate, 11, 39-40
Population, seasonal fluctuation in, 43
Population projections, 39-40, 43
Population pressure, 43-44
  final estimates, 45
  month-wise variations, 128-38
Power capacity, 59
Power cut, 87
  distribution, 83
Problem wards, 30
Projected water demand, 46-48
Prominant Institutions, 100
Property Taxation Committee, 18
Provincial services, 18
Public interventions, 22-24
Public problems, 22-23
Public Resource allocation, 11
Public schools, 37-38
Public urinals, 27
Public Works Committee, 18
Purkul Gaon power station, 59

Quasi-Government organisations, 12, 16, 17

Rainwater falling, 87
Residents' percaption, problems of, 22-24, 29, 39
Restaurants, 110

Rotary Club, 108

*Safai Karamcharis*, 26
Sanitation day, 106
Scrub-land, 64
Self-Help Environment Programme (SHEP), 105
Self-Sustaining project, 108
Separate bins, 102
Septic tanks, 90-91
Sewarage layout, 90-91
Sewarage system, 89
  areas of, 91-92
  proposed sewer line, 96
  zonal division of, 94
Slop categories, 60, 64, 72
  degree of, 69
  distribution of, 61-62
  landuse in, 65-66
State Government, grant in-aid and subsidy, 18-19
Storage tanks, 92
Supreme Court Monitoring Committee, 13
Surface traits, 52
Survey of India, Landour, 111
Sweepers, 101, 102

Tibetan Homes, 100, 111
  population, 37
Toll Tax collection, 100
Total landuse categories, 77
Tourism, 12
  current carrying capacity, 55-56
Tourism Department, 100
Tourists, accommodation for, 53, 55, 57-58, 83
Tourist flow per day, 54
Tourist in flow, 45, 107
  growth rate, 42
  hotel beds for, 45, 54
Tourist Pressure, 44-45
Traffic Jams, 55
Transport services, 27-28

Unemployment, 23
Urban development, 65
Urban eco-system, carrying capacity of, 35
Uttar Pradesh State Electricity Board, 25
Uttar Pradesh Urban Planning and Development Act, 12

Veterinary facilities, 27
Veterinary hospitals, 110

Waste disposal, 24, 39, 45, 109
  environmental implecations of, 99
  projected needs for, 53-54
Waste generation, 83
  estimates of, 49-52
  quantity of, 101
Waste management, 97-110
  by institution, 109-110
  current status of, 99-101
  effective methods of, 98
  problems, 83
  strategies, 107
Water requirement, 98
Water sources, 85-87
Water storage, 87-88
  supply, 48-49, 87-88
  scheme, 86
Water Supply and Sewerage Board, 35, 38, 85-96
Woodstock School, 105, 107, 109

Yellow bins, 102, 104

Zoning norms, 16